Engineering Anthropometry of the External Ear for Human-Product Interaction

Hao Fan • Dengkai Chen • Mengcheng Wang

Engineering Anthropometry of the External Ear for Human-Product Interaction

Methods and Applications of Physical Comfort Experience

Springer

Hao Fan
School of Mechanical Engineering
Northwestern Polytechnical University
Xi'an, Shaanxi, China

Dengkai Chen
School of Mechanical Engineering
Northwestern Polytechnical University
Xi'an, Shaanxi, China

Mengcheng Wang
School of Art and Design
Xi'an University of Technology
Xi'an, Shaanxi, China

ISBN 978-981-92-2159-2 ISBN 978-981-92-2160-8 (eBook)
https://doi.org/10.1007/978-981-92-2160-8

This work was supported by Key Laboratory of Industrial Design and Ergonomics, Ministry of Industry and Information Technology, China. This book was partly supported by Fundamental and Interdisciplinary Disciplines Breakthrough Plan of the Ministry of Education of China (Grant No. JYB2025XDXM605), and the Fundamental Research Funds for the Central Universities (NPU Grant No. D5000250035).

This Springer imprint is published by the registered company Springer Nature Singapore Pte Ltd.
The registered company address is: 152 Beach Road, #21-01/04 Gateway East, Singapore 189721, Singapore

Preface

The past two decades have witnessed the rapid proliferation of wearable technologies, which are now becoming deeply embedded in daily life, professional environments, and healthcare systems. Among these, ear-related wearables, which include hearing aids, communication headsets, earphones, earbuds, and emerging health-monitoring devices, have evolved into indispensable products for hundreds of millions of users. Yet behind this growth lies a persistent ergonomic challenge, which is that the human ear, particularly the auricle and the external acoustic meatus, is not simply another contact surface for technology but a highly individualized and sensitive anatomical structure. Designing for the ear therefore demands a unique anthropometric perspective, which connects biological diversity with engineering principles and user comfort requirements.

It is within this context that the present book, *Engineering Anthropometry of the External Ear for Human–Product Interaction: Methods and Applications of Physical Comfort Experience*, has taken shape. The purpose of this book is not only to document the anthropometric characteristics of the auricle and external acoustic meatus but also to explore how these measurements can be translated into predictive models of comfort, design guidelines, and practical tools for product developers. By drawing from ergonomics, human–machine interaction, and design engineering, the book positions itself as both a research monograph and a practical guide, which aims to bridge the gap between theoretical anthropometry and the pressing needs of wearable technology design.

The motivation for writing this book arises from two parallel developments. On one hand, wearable technology has become a central field of innovation, in which companies race to introduce devices that integrate communication, entertainment, health monitoring, and immersive experiences. On the other hand, dissatisfaction with in-ear devices remains common, which is often linked to discomfort, pain, or long-term usability issues. Unlike the hand, the head, or other parts of the body, the ear has received relatively limited attention in mainstream ergonomics. Most design processes still rely on simplified sizing assumptions or limited fit trials, which fail to capture the full variability of ear morphology across sex, age, and population groups. This gap is not merely an academic one, since it has direct consequences for user compliance, safety, and even health. An uncomfortable ear interface may lead users to abandon a device, which can compromise the accuracy of health sensors or increase the risk of ear canal injury. As wearables move deeper into healthcare and extended reality applications, the stakes are even higher, since comfort is no longer a luxury but a prerequisite for long-term adoption and reliable data collection. Addressing this challenge requires a systematic, data-driven, and human-centred framework, which integrates measurement, modelling, and application. This book is an attempt to provide such a framework.

First and foremost, this book seeks a balanced path in which theories and abstract models are distilled into applicable insights, while experimental methods and quantitative models are presented in ways that can be directly adopted by designers and engineers. The emphasis is not only on what the ear looks like or how it varies but also on how those variations influence the perception of physical comfort, and on how comfort thresholds can be translated into product design criteria.

Secondly, the book is not intended solely for ergonomics researchers, since it is written for a broad audience that includes wearable product designers, acoustic engineers, human factors specialists, healthcare professionals working with hearing technologies, and students of industrial design and biomedical engineering. By combining scientific rigor with practical orientation, the book aims to lower the threshold for applying ergonomic anthropometry to real-world design problems.

The book is organized into eight chapters, each of which contributes to the cumulative narrative that links ear morphology to product comfort and design. Chapter 1 introduces the foundations of engineering anthropometry, situating the external ear as a distinctive interface for wearable technology. It outlines the concepts of human–product interaction, comfort perception, and the role of anthropometry in ergonomic design. Chapter 2 presents the comfort factor model and the Ergonomic Discomfort-Comfort Factors (EDF) framework. This chapter explains how comfort is not a single dimension but a synthesis of mechanical, sensory, and cognitive factors, and how these can be experimentally extracted and quantified. Chapter 3 reports on an anthropometric analysis of the auricle and external acoustic meatus. In this chapter, 18 auricular variables and 22 meatus variables are defined, covering linear dimensions, angular parameters, circumferences, areas, and cross-sectional geometries. The analysis highlights persistent age-growth and phasic age-growth patterns, as well as sex differences, which together establish the empirical foundation for later modelling. Chapter 4 develops the deformation-comfort model, which introduces the concept of deformation ratio as a normalized index. It predicts subjective comfort ratings based on objective deformation data, defines comfort thresholds, and explains how they can be used to anticipate discomfort before actual testing. Rapid estimation methods are also proposed, allowing designers to apply predictive comfort modelling without the need for high-complexity simulation. Chapter 5 focuses on the auricle-meatus transitional structure and introduces the angle between tragus, entrance, and medial concha (ATEMC). It further defines the tragus expansion angle (TEA) and the tragus expansion index (TEI), establishing phasic age-growth patterns and linking transitional-structure expansion to comfort thresholds and exposure-response modelling. Chapter 6 examines the offset effects between functional-dimension perception and physical-dimension perception. Using Gaussian peak modelling of TEI-comfort relationships, it demonstrates a staged process of comfort perception in which functional stability can mask physical discomfort at certain expansion levels, and specifies sex-related exposure limits for intrusive wearables. Chapter 7 moves towards design mechanism modelling. It integrates anthropometric variability, deformation thresholds, and transitional-structure measures into a nested static model of ergonomic factors, while also incorporating self-reported comfort loss driven by usage motivation, scenario, and duration. Logistic E-R models of self-reported experience are presented as practical tools for predicting time-dependent comfort loss in daily contexts. Appendix provides supplementary resources, including methodological notes, demographic distributions, data tables, and percentile statistics of measurement variables. These materials serve as a reference for researchers and practitioners who wish to replicate or extend the methods.

If you are a product designer or engineer, you will benefit most from Chaps. 3–7, since these provide the anthropometric data, deformation-comfort thresholds, transitional-structure models, and design mechanism frameworks that are necessary for developing comfortable ear-related devices.

If you are an ergonomics or human factors researcher, you may wish to focus on Chaps. 2, 4, and 6, since these lay the conceptual and methodological foundations of comfort factor models, and deformation-comfort modelling.

If you are a healthcare professional or hearing technology specialist, Chaps. 3, 5, and 7 will be particularly relevant, since they demonstrate how anthropometric variability and transitional-structure expansion relate to comfort thresholds, exposure limits, and patient compliance in medical device usability.

For students or early-career researchers, the book as a whole offers a roadmap for integrating engineering anthropometry with ergonomics and product design. Reading sequentially from Chaps. 1 to 8 will provide a systematic overview, from theoretical foundations to applied design guidelines and supplementary resources.

The book originated from a series of research initiatives that began in 2015, when questions of ear comfort and anthropometry were still relatively underexplored. Over the course of more than a decade of research and practice, the project evolved through iterative measurement campaigns, modelling efforts, and collaborative discussions with colleagues in ergonomics, audiology, biomechanics, and industrial design. Each chapter represents not only a research outcome but also a stage in refining the methodological and theoretical framework, thereby embodying the cumulative insights gained over 10 years.

The writing of the book was supported by contributions from multiple collaborators, each of whom brought unique expertise in human factors engineering, statistical modelling, or design practice. Weekly discussions, which often extended late into the night, shaped the outline and ensured that the book would remain practical and comprehensible while still rigorous. The compilation process integrated diverse voices into a unified narrative, which reflects both the interdisciplinary nature of the subject and the long-term continuity of the research program.

Despite our best efforts to present a coherent and useful text, we acknowledge that there remain areas that require further refinement. Anthropometric variability across global populations, for instance, demands continued data collection and comparative analysis. Similarly, the predictive models developed in this book are only initial steps, which future refinements will need to extend by integrating dynamic conditions such as prolonged wear, physical activity, and environmental influences.

Nevertheless, the book aspires to establish a foundation for a new generation of ergonomic studies and product designs. By formalizing deformation ratio as a central metric, by introducing comfort thresholds, and by demonstrating rapid estimation methods, it seeks to transform how designers and researchers approach the interface between the human ear and wearable technology. We hope it will inspire further interdisciplinary collaborations and contribute to creating products that are not only functional but also genuinely comfortable for all users.

We extend our gratitude to the many colleagues, students, and participants who supported the data collection, modelling, and writing process. Their contributions have been invaluable in shaping the ideas presented here. We also welcome constructive criticism and suggestions from readers, since the field of ergonomic anthropometry is one that will continue to evolve with technological and societal change.

Finally, we would like to acknowledge the valuable support and help of the following researchers to the work related to this book over the past 10 years: Prof. Carisa Harris Adamson (University of California, San Francisco & Berkeley), Prof. Suihuai Yu (Northwestern Polytechnical University), Prof. Jianjie Chu (Northwestern Polytechnical University), Prof. Chunlei Chai (Zhejiang University), Prof. Olivier Doutres (École de Technologie Supérieure), Prof. Feng Gan (Southeast University), Dr. Jinlei Shi (Southwest Jiaotong University), Dr. Ao Jiang (Nanjing University of Aeronautics and Astronautics), and Dr. Xiao Zhao (Stanford University). This book is partly supported by Fundamental and Interdisciplinary Disciplines Breakthrough Plan of the Ministry of Education of China (Grant No. JYB2025XDXM605), and the Fundamental Research Funds for the Central Universities (NPU Grant No. D5000250035).

Xi'an, China
Oct. 2025

Hao Fan
Dengkai Chen
Mengcheng Wang

Contents

List of Figures

List of Tables

About the Authors

Hao Fan is an Associate Professor at Northwestern Polytechnical University and State Culture and Tourism Industry Think Tank—Chinese Art Evaluation Institute (Southeast University), where he directs the Wearable Ergonomics and Design Lab and serves as Assistant to the Director of the Key Laboratory of Ergonomics and Industrial Design, Ministry of Industry and Information Technology. He received his PhD degree from Northwestern Polytechnical University and completed a joint-PhD training program at the University of California Berkeley. Before returning to his current position, he held research and faculty appointments at Zhejiang University and Southeast University (China). Dr. Fan has engaged in external ear anthropometry research for more than 10 years, establishing a solid foundation in ergonomic measurement, modelling, and comfort evaluation. Dr. Fan's research focuses on ergonomic risk control and design strategies for wearable devices, emphasizing the integration of human–machine interaction, physiological assessment, and design translation. His work covers areas such as human–machine interface displays and alarm interactions, head–face wearable design, aircraft cabin interior ergonomics, and musculoskeletal disorders of the spine. He has published more than 60 academic papers and holds over 10 granted patents in these domains. Dr. Fan has led or contributed to over 20 research and industrial projects, with practical achievements widely applied in smart wearable product development and aviation ergonomic design, advancing both academic and industrial progress in the field.

Dengkai Chen is a Professor at Northwestern Polytechnical University (NPU). He received his PhD degree in Computer-Aided Industrial Design from Northwestern Polytechnical University. He currently serves as the Director of the Department of Industrial Design, the Key Laboratory of Industrial Design and Human–Machine Ergonomics, Ministry of Industry and Information Technology, and the Shaanxi Laboratory of Industrial Design Engineering. Professor Chen has led or participated in more than 10 national scientific research projects, including the National Science and Technology Support Program, the National Key R&D Special Project, the National Social Science Foundation, and the National Natural Science Foundation of China. He has presided over 10 major industrial design projects in key engineering fields such as manned spaceflight and manned deep-sea submersibles. He has published over 150 academic papers, authored seven books and textbooks, and holds more than 100 national patents. His primary research interests include product innovation design, design philosophy, technical aesthetics, and ergonomics design. He has completed numerous ergonomics research projects in manned spaceflight, aviation, and deep-sea operations, establishing a systematic ergonomics design and analysis framework. This framework covers diverse domains such as layout assessment and optimal design for confined compartments, visual ergonomics assessment, human factors reliability analysis in manned submersibles, dynamic risk evaluation for complex operational systems, performance assessment of special operators, and comprehensive evaluation of operational comfort and maintainability. His research further extends to illumination optimization in confined environments, operator environment design for performance enhancement, and AI-driven design innovation.

Mengcheng Wang is an Assistant Professor in the Department of Industrial Design, Xi'an University of Technology. She received her PhD degree from Northwestern Polytechnical University and completed a joint-PhD program at the School of Public Health, University of California Berkeley. Dr Wang has been engaged in ergonomic risk control and design optimization research for wearable devices, with more than 10 years of experience in external ear anthropometry and wearable comfort modeling. She has led or participated in ergonomics research and design projects for leading international companies, including Tesla, Liberty Mutual, Google, Huawei, and AVIC, bridging academic research with industrial innovation. Her research focuses on human-machine interaction, biomechanical modeling, and multimodal comfort evaluation in wearable systems. Her studies have advanced the understanding of auricular and external acoustic meatus morphology, in-ear comfort mechanisms, and upper-limb muscle activity modelling. Dr Wang's interdisciplinary research integrates neural and physiological signal analysis, machine learning, and ergonomic design methodologies, contributing to the development of human-centred design frameworks for smart wearables and interactive products.

Abbreviations

3D	Three-dimensional
ACE	Azimuth angle of the geometric centre of the entrance
AIA	Auricular inclination angle
AL	Auricular length
ANOVA	Analysis of variance
AS-I	Azimuth angle of Section I
AS-II	Azimuth angle of Section II
ATEMC	Angle between tragus, entrance, and medial concha
AWI	Auricular width at the incisura anterior auris posterior
AWP	Auricular width at postaurale
BMI	Body mass index
CAO	Concho-mastoid angle at otobasion posterius
CAP	Concho-mastoid angle at postaurale
CCL	Cavum concha length
CCW	Cavum concha width
CD	Conchal depth
CL	Conchal length
CT-based	Computed tomography based
CV	Coefficient of variation
CW	Conchal width
DRexps	Deformation ratio (expansion)
DRext	Deformation ratio (extension)
EA	Entrance area
EAM	External acoustic meatus
EC	Entrance circumference
ED-C	Entrance depth from geometric centre to medial concha
EDFs	Ergonomic design factors
ED-P	Entrance depth at posterior from medial concha
EL	Entrance length
EML	Ear morphologic length
EPS	Ear protrusion at the superaurale level

EPT	Ear protrusion at the tragal level
ES-I	Elevation angle of Section I
ES-II	Elevation angle of Section II
EW	Entrance width
FASVs	Factors associated with shape variations
FBA	First bend area
FBC	First bend circumference
FBL	First bend length
FBW	First bend width
HSD test	Tukey's honest significant difference test
KMO	Kaiser–Meyer–Olkin test
LL	Lobular length
LW	Lobular width
PCA	Principal component analysis
SBA	Second bend area
SBC	Second bend circumference
SBL	Second bend length
SBW	Second bend width
SD	Standard deviation
SD-I	Section I depth
SD-II-E	Section II depth from entrance
SD-II-F	Section II depth from first bend
S-W test	Shapiro–Wilk test
TEI	Tragus expansion index
TH	Tragal height
TL	Tragal length

Chapter 1
Introduction to Human-Product Interaction, Comfort Perception, and Ear Anthropometry

Abstract Human-cantered design has become a central paradigm in contemporary engineering and product development, driven by advances in computation, sensing, and wearable technologies. Comfort, as a key determinant of user experience, plays a critical role in evaluating and optimizing human-product interaction. Ear-related wearable devices represent a rapidly expanding category within head-face wearables, evolving from simple audio interfaces into multifunctional systems that support communication, noise regulation, hearing assistance, health monitoring, and biometric sensing. Despite their widespread use, ergonomic research focusing specifically on the external ear remains limited, creating challenges for comfort-oriented design and evaluation. Human-product interaction is shaped by the combined influence of user characteristics, product attributes, and interaction processes. Established interaction experience models provide complementary perspectives for understanding how perception, cognition, and physical exposure contribute to user outcomes. Comfort is inherently multidimensional, encompassing physical, psychological, physiological, and environmental components, with physical comfort serving as the most direct and measurable indicator in wearable contexts. Ear anthropometry offers essential quantitative foundations for ergonomic design by capturing variability in ear size, shape, and population characteristics. Together, these perspectives establish a theoretical and methodological framework for analysing comfort mechanisms and supporting evidence-based design of ear-worn wearable technologies.

Keywords Human-product interaction · Comfort perception · Physical ergonomics · Ear-related wearable devices · Ear anthropometry

Over the past decade, engineering has undergone a profound transformation in design and manufacturing, driven by advances in computing, sensing, and wireless communication technologies. Developments in design and manufacturing have accelerated the integration of human-centred design with large-scale manufacturing systems, shifting research and practice toward approaches that embed human factors and ergonomics across the entire product lifecycle. Within the developments of design and manufacturing, product design is no longer limited to aesthetics or

H. Fan et al., *Engineering Anthropometry of the External Ear for Human-Product Interaction*, https://doi.org/10.1007/978-981-92-2160-8_1

function but serves as a central hub that connects humans, design, and manufacturing in an intelligent and interactive network.

A key dimension of the transformation is the growing emphasis on user comfort, which serves both as a determinant of product experience and as a competitive advantage for enterprises. Research on comfort perception and its determinants has expanded rapidly across sectors such as aircraft interiors [1, 2], cabin seating [3, 4], biomechanical exoskeletons [5, 6], upper-limb and distal wearables [7–9], head-face wearables [10, 11], and equipment for individuals with higher body mass indices [12–14]. Among these, ear-related wearables have emerged as a distinct category within head-face wearable systems. Once designed primarily for sound transmission or isolation, ear-related wearables have evolved into multifunctional interactive tools providing communication, noise control, hearing assistance, health monitoring, bio-signal acquisition, and location tracking [15–17] (Fig. 1.1). They are now indispensable across domains such as aviation, aerospace, transportation, protection and healthcare [18, 19].

Despite the ubiquity of ear-related wearables, studies of ear ergonomics and comfort remain relatively underdeveloped compared to other wearable systems. Long treated as a subset of head-face devices, the external ear has received limited attention as an independent locus of ergonomic exposure. Yet, comfort outcomes are highly population-specific and are influenced by variations in body dimensions, musculoskeletal attributes, and skin biomechanics across regions and demographics. Addressing these gaps requires systematic investigation of ear anthropometry and ergonomic factors to establish evidence-based design principles for ear-related wearables.

Comfort itself is multidimensional, integrating physical, psychological, physiological, and environmental dimensions. Among these, physical comfort is the perceptual response to ergonomic factors during direct human-product interaction and

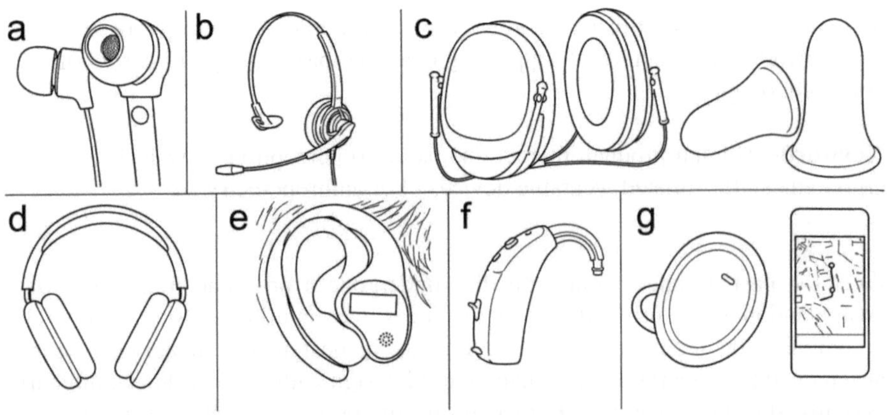

Fig. 1.1 Ear-related wearable devices categorized by functional mechanisms: (**a**) real-time voice transmission; (**b**) instant communication; (**c**) hearing protection; (**d**) environmental noise regulation; (**e**) voice-based interaction; (**f**) hearing aids; (**g**) geo-positioning

is the most immediate and measurable indicator of fit and usability [20, 21]. For wearable devices, issues of physical comfort stem directly from the act of wearing, and optimizing this experience has become a priority in both research and practice.

Ultimately, comfort is both a core aspect of user experience (e.g., trust, satisfaction, convenience) and a central theme of ergonomics (safety, health, comfort, efficiency). Establishing quantitative relationships between ergonomic exposures and user outcomes is therefore essential. Such efforts can provide standardized metrics, design guidelines, and evaluation tools that advance safety, health, and efficiency while supporting the growing role of wearable technologies (for ear-related wearables) in daily life and professional applications [22].

1.1 Human-Product Interaction Experience

1.1.1 General Structure and Perspectives

Because humans possess multiple biological systems that are independent of the external environment (such as the motor system, cognitive system, and sensory system), we are able to interact with the environment outside the body. Products constitute an important part of this external environment. Direct physical contact or interaction between users and products triggers active instrumental activities (e.g., wearing work clothes, putting on a hat), passive perceptual feedback (e.g., whether it feels comfortable), and cognitive reflections (e.g., adjusting usage methods or replacing the product). Based on this perspective, Clarkson [23] defined the general process of human-product interaction experience as a circular structure (Fig. 1.2). Schifferstein and Hekkert [22] further proposed that there is an interaction between the user's experiential outcomes and the process of use. Accordingly, they redefined the general process of human-product interaction experience as a dual-cue structure (Fig. 1.3). In the Schifferstein-Hekkert model, the frontend represents the users and their perception of information. The users provide the necessary drive, sensitivity, cognitive capacity, and attentional capability for interaction through their motor,

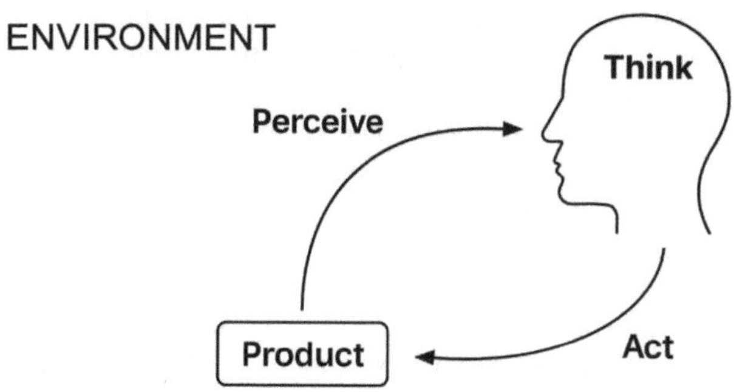

Fig. 1.2 The Clarkson model [23] of human-product interaction experience

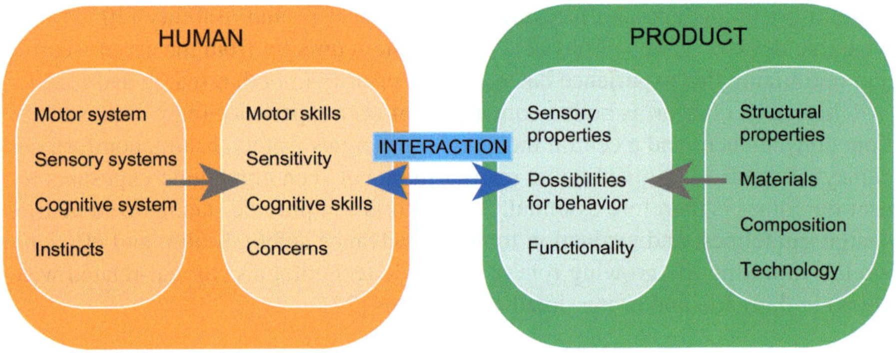

Fig. 1.3 The Schifferstein-Hekkert model [22] of human-product interaction experience

sensory, and cognitive systems as well as instincts. The backend of Schifferstein-Hekkert model represents the product and its associated information. The product contributes to the interaction through its structural attributes, materials, composition, technologies, and labels, thereby supplying the perceptual qualities, behavioural affordances, and functional properties required for interaction. Based on this, user experience research could be divided into three perspectives according to the primary research subject:

1. User experience perspective: this perspective focuses on the user, primarily studying user attributes or behavioural characteristics.
2. Product perspective: this perspective focuses on the product itself, typically involving case studies of specific product types or parameters.
3. Interaction perspective: this perspective focuses on interaction structure, process, or pattern, generally involving analysis and research of specific interaction structures or methods.

1.1.2 Interaction Experience Models

Depending on the research perspective, models of human-product interaction experience could generally be categorized into three types: frontend perspective models, backend perspective models, and interaction perspective models. The fundamental components of these models include frontend (the user, as a primary component), backend (the product, as a primary component), information chains (with descriptive elements), and secondary components connected by the information chains (e.g., user attributes, product attributes, interaction modes, and usage motives).

Models of human-product interaction experience can be distinguished according to their analytical perspective, each emphasizing different focal elements of the interaction process. Frontend perspective models highlight the diversity of users and aim to provide differentiated feedback based on user attributes. Backend perspective models focus on product variations, examining how functional differentiation across product types shapes user experience and societal impact. Interaction perspective models, in contrast, emphasize the dynamics of motivation and cognitive processes during use, offering insights into how motives and individualized perceptions shape interactive experiences. Together, these three perspectives provide complementary frameworks for understanding and optimizing human-product interaction.

In summary, the three analytical perspectives, frontend, backend, and interaction, each illuminate distinct yet interrelated dimensions of human-product interaction. The frontend perspective underscores user diversity, the backend perspective highlights product differentiation, and the interaction perspective emphasizes motivational and cognitive dynamics. Taken together, these perspectives provide a complementary and integrative framework, offering a solid theoretical foundation for subsequent discussions on how to evaluate, model and ultimately optimize human-product interaction.

In work systems or public entertainment systems, a single product or device often needs to provide both general and user-specific functions or services for a large number of users. Consequently, it is necessary to deliver differentiated usability feedback for multi-user scenarios based on individual user attributes or behavioural capabilities. Building on Clarkson model, Chang et al. [24] proposed the (multi-)frontend model. Compared with general process models, the (multi-)frontend model aims to provide differentiated and targeted feedback mechanisms for each individual user within a single product or device, thereby optimizing the overall user experience of the group.

Due to the differentiated design of specific functions, many products often integrate numerous subtypes (e.g., automobiles, watches). Considering that functional differences may lead to divergent user experiences, Fokkinga et al. [25] proposed that interaction experience models should emphasize the experiential relationship between users and different product subtypes, thereby facilitating a broader understanding of the design significance of related products (Fig. 1.4). In the model proposed by Fokkinga et al., the information chain follows a bottom-up progression: beginning with product attributes and human-product interaction, followed by the user experience and feedback associated with different product types, and culminating in the impact of products on social life.

For products with complex functional options or operational processes, Demir et al. [26] suggested that understanding user motives can help define the interaction structure, process, or mode in a way that optimizes the human-product interaction experience. Accordingly, in the motive hierarchy model, Demir et al. defined three levels of motive identification (general motives, contextualized motives, and interaction goals), together with corresponding motivational components This model establishes information chains linking lower- and higher-level motives by identifying motive consistency components and their respective hierarchical levels, thereby determining reasonable pathways for improving interaction experience. In addition, Walls [27] proposed that individual users pursue unique experiences in human-product interactions. Therefore, interaction structures, processes, or modes tailored

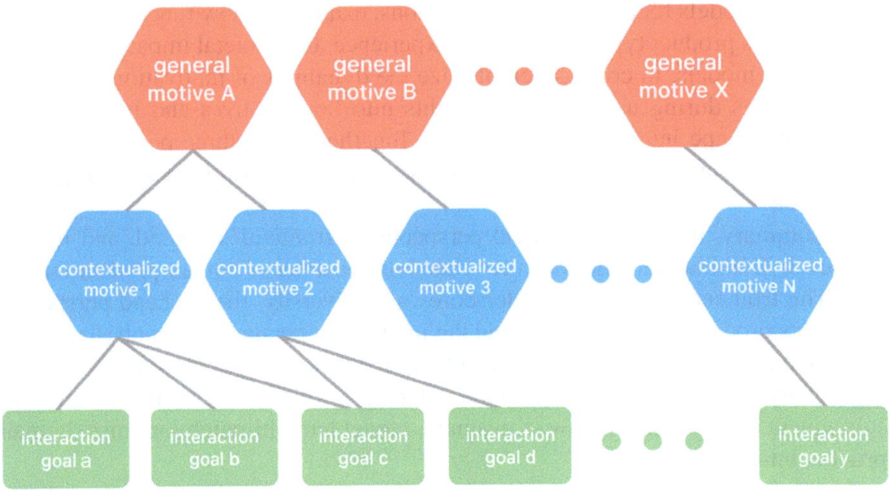

Fig. 1.4 The motive hierarchy model [26] of human-product interaction experience

to individuals should integrate perceptual processes, cognition, association, and memory.

1.2 Comfort and Physical Ergonomic Factors

The levels of human health, safety, and efficiency fluctuate with changes in users' perceived comfort feedback. The primary subject of comfort is the human (frontend), while the focus of investigation lies in tools, equipment, and task systems (backend). Based on the classification of backend content, common types of human physical comfort perception include operational comfort perception, wearable comfort perception, and seating comfort perception.

1.2.1 Comfort Perception

Perceptual Process: Comfort and Discomfort

Comfort perception integrates both comfort and discomfort as experienced by the user. Comfort represents a perceived state of ease and balance. Discomfort, in addition to being a subjective expression of the user's perception, also serves as one of the predictive indicators of health or safety risks. Helander and Zhang [28] proposed that both comfort and discomfort should be simultaneously considered in the process of comfort perception. In quantitative studies of comfort perception, De Looze

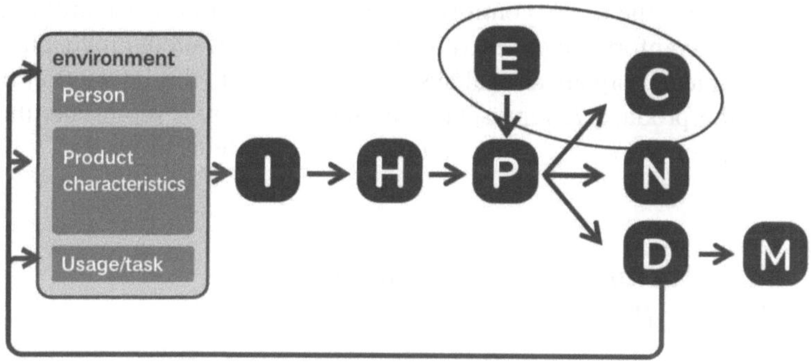

Fig. 1.5 The Vink-Hallbeck perception model [20] of (dis)comfort

et al. [29] suggested that discomfort is more directly linked to objectively measurable doses of Physical Ergonomic Exposure.

Vink and Hallbeck [20] proposed that comfort and discomfort originate from the interaction (I) between products (P) and persons (P) when performing specific activities (U) within a given environment. This interaction (I) leads to changes in human effects (H), such as alterations in sensory receptors, tactile feedback, body posture, and muscle activation. Perceptual feedback (P) is influenced not only by variations in human attributes but also by interaction expectations (E). Discomfort, in turn, may result in skin tissue and musculoskeletal injuries (M) (Fig. 1.5).

Types of Comfort Perception

Comfort issues reflect the state of interaction between humans and their external environment in terms of physical, psychological/cognitive, physiological, and environmental/organizational dimensions.

Physical comfort perception arises from the comfort and/or discomfort sensations generated by changes in Physical Ergonomic Exposure. Psychological/cognitive comfort perception results from variations in visual stimulation, psychological activity levels, and cognitive load. Physiological comfort perception stems from changes in physiological indicators such as body temperature, heart rate, and blood pressure. Environmental/organizational/system comfort perception is shaped by changes in environmental, organizational, or systemic structures or parameters (e.g., temperature, humidity, personal space, organizational density). Based on the comfort perception process, Naddeo et al. [30] proposed the Naddeo-Capetti conceptual equations, which distinguish among different types of comfort:

$$C_i = P_i\left(f\left(I,H\right)\right) - E_i \tag{1.1}$$

$$D_i = P_i\left(g\left(I,H\right)\right) + E_i \tag{1.2}$$

where, C represents the user's comfort, D represents the user's discomfort, i represents the type of comfort perception (physical, psychological/cognitive, physiological, or environmental/organizational/system), I represents the interaction between the user and the product, H represents the human effects, P represents the user's comfort perception, E represents the interaction expectations or losses.

1.2.2 Basic Factors Associated with Wearing Comfort Perception

Identifying basic factors and constructing research indicator models is the first step in comfort design and research. The conventional approach involves conducting a literature review and field survey to design a questionnaire; recruiting participants to complete the questionnaire and rate the importance or impact of each item; and finally, performing statistical analysis (such as difference analysis or dimensionality reduction) to determine the basic factors.

Questionnaire (survey) is a fundamental method and data collection approach in physical ergonomic exposure studies, which qualitatively captures participants' actual preferences and key concerns, offering practical references for theoretical model applications. Questionnaire (survey) also enables the collection of participant feedback on the research topic, thereby providing empirical foundations for factor extraction and quantitative analysis. Common scoring scales include percentage scales, (±)5-point Likert scales, and 10-point scales. Dimensionality reduction methods include principal component analysis (PCA), factor analysis, and cluster analysis. Among these methods, factor analysis has proven effective for feature extraction and factor identification in engineering design, biometrics, engineering anthropometry, and ergonomic design, especially in identifying associations among human attribute variations and physical ergonomic exposure factors. Currently, wearable comfort perception (i.e., comfort associated with human wearable devices) focuses on devices applied to the head-face system (respirators, headgear, eyewear, etc.), the limb system (shoes, gloves, wristbands, joint protectors, exoskeletons, etc.), the axial system (chest/neck supports, lumbar braces, spinal exoskeletons), the external ear (headphones, hearing aids, hearing protectors), and the whole body (workwear, protective clothing, industrial exoskeletons).

Basic factors associated with wearable comfort include duration of use (time load), product dimensions, and skin surface-pressure sensitivity. For devices targeting the musculoskeletal or axial systems (spine, limbs), additional basic factors include joint flexibility and extent of body/skin coverage. Since duration of use is a fundamental factor contributing to human fatigue, it is typically one of the core factors considered in all studies related to comfort. User comfort perception feedback often follows a trajectory of "adaptation → (temporary) stability → decline" as time load increases. For hearing protection devices, the known basic factors associated with wearable comfort are design dimensions, skin surface-pressure sensitivity, and duration of use. Due to the elasticity of auricular cartilage, the comfort-fit size of devices is often slightly larger than the actual anatomical size of users [31, 32].

1.2.3 Physical Ergonomic Factors

In activities of large-scale production, manufacturing, and collaborative design, ergonomics serves a crucial bridging role. Based on the differences in research subject matter, ergonomics can be divided into physical ergonomics (with a focus on the human body), cognitive ergonomics (with a focus on vision, brain, and mental activities), and macro-ergonomics (with a focus on systems, organizations, and environments related to humans).

Ergonomic factors are people-oriented and focus on factors that constrain or influence users' safety, health, comfort, and efficiency during daily work, life, or the execution of specific tasks. It seeks to explain exposure mechanisms by establishing quantitative relationships or effect linkages between outcomes (such as safety, health, comfort, and efficiency) and factors [33, 34]. Ergonomic factors generally originate from users' physical, cognitive/psychological attributes, as well as from tools, systems, organizations, and environments with which they interact. Physical ergonomic factors mainly concern human body dimensions and the biomechanical properties of the musculoskeletal system and skin. These are influenced by age, height, weight, and gender. One of the primary goals of physical ergonomic factors research is to establish balanced or win-win relationships between humans and objects (e.g., tools, tasks), thereby providing recommendations to reduce user risk and offering theoretical support and design criteria for ergonomics design activities. From the perspective of human-product interaction, research on physical ergonomic factors typically follows two pathways: frontend and backend.

Frontend pathway involves analysing users' physical ergonomic factors to draw conclusions about whether current designs are favourable or unfavourable, or to provide references for potential/future design content or formats. The goal of frontend pathway is to produce broadly applicable design recommendations, guidelines, or theories. Whether product design parameters match users is a critical determinant of perceived wearable comfort. Choosing the frontend pathway in physical ergonomic factors research for wearable comfort could provide usable thresholds and ranges of ergonomic factors for parameter design of wearable devices, i.e., the endurance range of (dis)comfort. *Example*: Harris et al. [35, 36] conducted a 6.5-year cumulative study on 2474 workers' tool exertion and found that when grip and pinch forces were below 45 N and 10 N, respectively, both hand discomfort and musculoskeletal disorder risks during tool use were significantly reduced. Backend pathway starts from a predetermined design target (usually a specific product model) and investigates how variations in design parameters influence comfort. The outcomes are highly targeted design recommendations, guidelines, or theories. *Example*: Li et al. [37] and Zhao et al. [38] studied economy-class aircraft seating with different seat pitches (28–36 inches) and found that 28-inch spacing yielded the worst long-duration sitting comfort, while 30-inch spacing performed best overall.

1.3 Ear Anthropometry for Ergonomics

Anthropometry, the scientific discipline concerned with the measurement of human body dimensions, has long served as a cornerstone of ergonomics, medicine, and industrial design. Among the many anatomical regions studied, the external ear (also known as the auricle or pinna) has historically received less attention compared to other body parts such as the head, hand, or torso. However, in recent decades, ear anthropometry has emerged as a specialized subfield of human measurement, driven by the increasing importance of ear-related applications in medical diagnostics, hearing aid development, forensic science, personal protective equipment, and wearable device design.

The external ear plays a critical role in auditory perception by shaping sound localization cues and amplifying certain frequency ranges. Beyond its auditory function, the ear serves as a key anatomical landmark for surgical procedures, a distinctive biometric trait for personal identification, and a sensitive interface for wearable products such as headphones, earphones, and augmented reality devices. Understanding the variability of ear size, shape, and surface characteristics is therefore essential for improving both scientific knowledge and technological innovation.

Early anthropological studies in the nineteenth and twentieth centuries occasionally documented ear dimensions as part of broader craniofacial research. These efforts were largely descriptive, aiming to classify populations or explore genetic inheritance patterns. During the mid-twentieth century, ear measurements became increasingly relevant for forensic science, as investigators recognized the uniqueness of auricular morphology. In the late twentieth century, advancements in audiology and the growing demand for hearing aids stimulated research into ear canal dimensions and auricular shape, as these factors influence acoustic coupling and device comfort. The rise of ergonomics and product design in the 1990s and 2000s further expanded the need for precise anthropometric data, particularly with the proliferation of personal audio products and head-mounted displays. Today, ear anthropometry is an interdisciplinary field that integrates engineering anthropometry, medical imaging, computational modelling, and biometric technologies.

1.3.1 Measurement Methods

Engineering anthropometry is regarded as one of the core methodological foundations for ergonomics research aimed at "fitting tasks and tools to humans". In 1968, Alexander et al. [39] conducted an engineering anthropometric study of the external ear of Air Force pilots using photogrammetric techniques, which provided a fundamental database and methodological reference for subsequent analyses of ear anthropometry and comfort-related design studies.

Data Collection Methods

Depending on the measuring instruments employed, anthropometric data acquisition methods for the external ear can be broadly categorized into direct/contact methods and indirect/non-contact methods (Fig. 1.6).

Direct/contact methods are widely applied due to their low cost and operational efficiency. During data collection, water-soluble markers or adhesive stickers are typically used to indicate anatomical landmarks or structural turning points on the subject's skin. Observers then perform repeated measurements for each variable to calculate systematic errors or assess inter-observer differences. The most commonly used tools, such as callipers and attached goniometers, are similar to those employed in anatomy and forensic science. However, certain risks exist, including potential allergic reactions to marking materials or minor skin abrasions caused by measuring instruments.

Indirect/non-contact methods include photogrammetry, conventional three-dimensional scanning (laser or structured light scanning), and computed tomography based (CT-based) three-dimensional reconstruction. Photogrammetric accuracy largely depends on the resolution of the imaging device and the measurement grid. Conventional 3D scanning and CT-based reconstruction minimize measurement errors associated with skin deformation and parallax distortion, thereby overcoming the precision limitations of direct and photographic methods. CT-based reconstruction, in particular, has enabled comprehensive analyses of auricular variables, such as the earlobe, auricle, tragus, and antitragus, by quantifying both linear and angular parameters. Nevertheless, potential radiation exposure restricts application of CT-based reconstruction in large-scale population studies. Thus, CT-based reconstruction is largely limited to clinical or research contexts rather than large-scale ergonomic surveys.

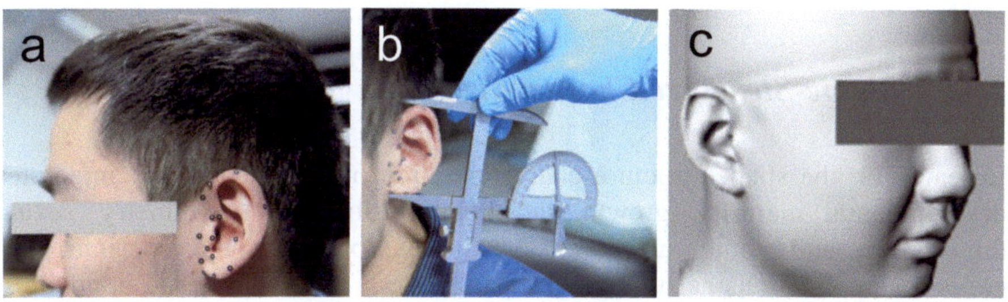

Fig. 1.6 Methods of measurement and data collection for the external ear: (**a**) landmarks; (**b**) direct measurement; (**c**) three-dimensional scanning

Measurement Variables

With the continuous development of ergonomics design, clinical diagnosis, and personal identification, the number of variables required for the shape-structure characterization of the external ear has steadily increased. Earlier studies (prior to 2000) focused narrowly on specific measurement variables, such as auricular length and width. More recent investigations, however, emphasize overall auricular morphology to enhance the generalizability of findings beyond a limited set of anatomical structures.

Unlike the relatively underexplored field of the external acoustic meatus, ear anthropometry has developed substantially over the past three decades. At present, more than 15 key auricular measurement variables have been identified. Beyond traditional measures such as length and width, measurement variables related to the earlobe, concha, auricular protrusion, and auriculo-mastoid angle have been progressively incorporated into study frameworks [40, 41]. The continuing proposal of novel measurement variables and measurement methods has facilitated more comprehensive shape-structure characterization. However, given that most studies are tailored to specific product models, the systemic integration and interrelation of measurement variables remain limited. For instance, Purkait et al. [42] examined auricular variation across age groups in 415 Indian males aged over 18, defining nine measurement variables and discussing earlobe degeneration for the first time. Wang et al. [43] analysed auricular dimensions in 485 northern Chinese adults, expanding the variable set to twelve. Subsequently, Alexander et al. [44] performed a systematic analysis of auricular morphology in 420 subjects aged over 15 from four ethnic groups, further expanding the variable set to thirteen and, drawing on Ferrario's perspective [40], redefining auricular orientation through the construction of an ear axis. Recently, Lee et al. [45] and Fan et al. [14] each defined over fifteen key variables and proposed fitting intervals suitable for auricular comfort in wearable design. Collectively, these efforts have established a stable framework comprising thirteen core variables plus targeted supplementary variables, which reflects both the increasing refinement of research on auricular comfort and the growing prevalence of fit-related challenges in product applications.

1.3.2 Shape-Structure Characterization

Research on the shape-structure characterization of the external ear integrates developmental maturity, growth, degeneration, bilateral asymmetry, sexual dimorphism, and ethnic variation. Such findings provide critical design references for applications across age groups, genders, individuals, and ethnicities. Currently, the auricular characterization studies are relatively comprehensive and advanced.

Growth Trajectory

According to Purkait [46] and Standring [47], the developmental maturity of the auricle occurs between 5 and 16 years, with complete growth achieved by approximately age 16. Length-related variables mature more slowly than width-related ones, although width reaches maturity earlier. Males generally attain auricular maturity later than females, with auricular length, earlobe length, and conchal length being the latest to stabilize. Zhao et al. [48], in a study of the Chinese population, reported that auricular length reaches maturity at approximately age 14, representing the latest-maturing variable. Importantly, auricular shape continues to change after adulthood. Most anatomical structures of human typically follow a trajectory of "growth → maturity → stability → degeneration" (e.g., height, spinal length). However, auricular variables often exhibit a "growth → maturity → continued growth" trajectory, including length and width. Degeneration of auricular shape does occur, particularly among older populations, with variables such as earlobe width, auricular protrusion, and auriculo-mastoid angle being especially affected. Other variables may also degenerate, but current evidence remains insufficient.

Bilateral and Sexual Differences

Bilateral symmetry of the auricle is a key ergonomic consideration, underpinning paired design in wearable devices. Earlier studies (prior to 2005) reported high bilateral symmetry in auricular shape and dimensions across multiple populations. These conclusions, however, were based on limited key variables (e.g., length, width). With the expansion of measurement variable sets, findings have been revised: auricular dimensions remain broadly symmetrical, but significant bilateral differences are observed in certain key variables.

For example, Purkait et al. [42], Ahmed [49], and Alexander [44] demonstrated population- and sex-specific patterns of bilateral differences. Among Sudanese populations, auricular length exhibited bilateral asymmetry, while auricular width differences were significant only in Sudanese females. In central Indian males, both auricular length and width showed significant asymmetry. Regarding sex differences, most studies report that male auricles are larger than female auricles, with many variables showing significant sexual dimorphism.

Ethnic Differences

Ethnic variation also plays a critical role in auricular shape. For instance, European populations generally have larger auricles than South Asian populations (e.g., Indians), whereas Korean populations exhibit larger auricles than Europeans. Wang et al. [43] have published data on northern Chinese Han populations. Although China's north-south division can be drawn along the Qinling-Huaihe Line (geography) or the Hu Huanyong Line (population density), Wang et al. did not specify the geographic origins of their participants, suggesting that their findings only partially reflect regional auricular characteristics.

Age Grouping and Age Limits

Age-based grouping in shape-structure studies typically follows intervals of 10, 15, or 20 years. For ergonomic design applications, the presence of significant intergroup differences is essential for age-stratified design. In the absence of detailed growth trajectory data, a 10-year interval is often adopted. The elderly population represents an increasingly important focus in ergonomics research. Historically, upper age limits in auricular studies have been capped at 60 years, largely due to average life expectancy constraints and challenges in sampling older adults.

1.3.3 Challenges in Ear Anthropometry

Despite the substantial progress made in ear anthropometry, several challenges continue to limit the field's practical application. At the same time, emerging technologies and evolving design need to create significant opportunities for innovation.

Historical Context

Late 1960s to 1990s

According to available research records, the Aerospace Medical Research Laboratories of the U.S. Air Force and the National Aeronautics and Space Administration (NASA) were among the earliest institutions to conduct exploratory studies on auricular wearing comfort [39, 50]. During this period, the research primarily focused on defining the applicable structures and components of the auricle, proposing analytical indicators for engineering anthropometry (measurement variables and measurement orientations), and establishing basic methods for data acquisition and analysis. The outcomes were mainly applied to safety analysis of mission processes and reliability verification of equipment design, marking the initial stage of wearing comfort research within the context of ergonomics and aerospace applications.

2000 to 2010

With the gradual refinement of comfort theory and the increasingly independent development trend of auricular wearable devices, countries such as Europe, Japan, and South Korea launched systematic studies on auricular wearing comfort. During this stage, more comprehensive theoretical discussions emerged, covering research objectives, types of comfort perception, types of ergonomic exposure, and product classification. The research focus gradually shifted from static morphological analysis of the auricle to ergonomic evaluations of auricular fitting structures. Due to the explosive growth of product forms and functional categories, most studies adopted back-end approaches for experimental validation and method testing. Although their findings exhibited certain limitations in generalizability and extensibility, they provided a solid theoretical and methodological foundation for subsequent integrative and interdisciplinary research.

2010 to Present

Since 2010, research on ear wearing comfort has increasingly emphasized group differences and individualized design. Researchers began to focus on intra-population ergonomic factors, employing large-scale sampling and data analysis to establish comfort design standards applicable to specific national or regional populations. Meanwhile, the research scope gradually extended from the auricle to the external auditory meatus, though controversial issues have become more pronounced. First, despite advances in defining structural dimensions along the depth direction, fitting structures beyond the auricular orifice remain a research blind spot. Second, auricular components associated with wearing comfort have not yet been systematically integrated. In terms of methodological orientation, front-end and back-end approaches in general studies continue to lack effective linkage. Regarding ergonomic factors, studies still primarily rely on classical linear dimensions, with insufficient morphological characterization and an underdeveloped definition of auricular fitting intervals. Furthermore, due to the absence of effective exposure quantification models and research frameworks, potential design risks cannot be adequately assessed in the early stages of design planning.

Methodological Standardization

A critical challenge in ear anthropometry lies in the lack of universally standardized measurement protocols and reference frameworks. Although a wide range of direct and indirect measurement techniques has been developed, from traditional callipers to state-of-the-art three-dimensional imaging systems, comparability across studies remains limited. Variations in landmark definition, scanning resolution, and error-correction protocols have led to inconsistent datasets, undermining the reliability of cross-study comparisons. For instance, the choice of anatomical landmarks on the auricle may vary subtly across investigators, producing significant discrepancies when aggregated at scale. Similarly, the calibration of photogrammetric systems or parameters for structured-light scanning can produce systematic biases that are difficult to reconcile without standard reference models.

Another methodological limitation concerns the dynamic nature of auricular morphology. Unlike skeletal anthropometry, the auricle is composed largely of elastic cartilage covered by soft tissue, making it highly susceptible to deformation from contact or postural changes. Direct measurement methods, while inexpensive, often induce such deformations, whereas non-contact methods mitigate them but introduce challenges of resolution and computational reconstruction.

Emerging digital technologies provide new opportunities. Advances in high-resolution 3D imaging, combined with machine learning-based feature extraction, promise to standardize measurement pipelines. Automated landmark recognition algorithms can reduce observer-dependent variability, while cloud-based reference databases may enable cross-study harmonization. Furthermore, integration with biomechanical modelling platforms could provide correction factors for tissue deformation, improving both accuracy and reproducibility.

Population Diversity and Representativeness

Another major challenge in ear anthropometry is the insufficient representativeness of existing datasets. Most large-scale studies are concentrated within limited geographic regions or specific ethnic groups, leading to biased reference standards. Even within a single country, regional variation can be substantial. In China, for instance, the lack of data spanning both northern and southern populations have restricted the capacity to generalize findings across the entire population. Such limitations impede the ability to establish truly universal ergonomic design standards or databases. Elderly populations, despite being increasingly relevant for ergonomic and medical applications, remain understudied due to difficulties in recruitment and measurement. Gender balance is another concern, some datasets are heavily skewed toward male participants, constraining the validity of sex-specific conclusions.

Opportunities lie in the integration of large-scale, multi-ethnic population studies facilitated by digital health infrastructure. The growing adoption of wearable devices and mobile scanning applications offers unprecedented avenues for data collection beyond laboratory environments. Crowd-sourced anthropometric datasets, if combined with rigorous data curation, could capture greater variability across age, sex, and ethnic dimensions. Such initiatives would not only enhance the scientific validity of ear anthropometry but also support inclusive design in global markets.

Integration into Ergonomic Applications

While ear anthropometry has achieved technical maturity, its translation into practical applications presents both challenges and opportunities. In ergonomic design, one persistent challenge is the limited incorporation of auricular data into product development pipelines. Many wearable devices, such as earphones, hearing aids, and protective equipment, are still designed using limited anthropometric references, often derived from small or outdated datasets. This results in suboptimal fitness, discomfort, and even health risks such as pressure-induced dermatitis or acoustic leakage. Bridging the gap between academic anthropometric research and industrial design practice remains a pressing challenge.

Opportunities are emerging through interdisciplinary integration. In ergonomics, the convergence of anthropometry, computational simulation, and user-centred design enables predictive modelling of comfort and usability across diverse populations. Parametric design tools, informed by large-scale anthropometric datasets, could facilitate customizable product development, ensuring that devices adapt to individual auricular geometries. Ultimately, the successful translation of ear anthropometry into practice depends on stronger collaborations across disciplines. Strong partnerships between researchers, industry designers, and clinicians are essential to align measurement standards, validate application protocols, and promote inclusive and ethical use of anthropometric data.

Competing Interests This study is by "the Fundamental Research Funds for the Central Universities" (NPU Grant No. D5000250035, SEU Grant No. 2242024S30013). Additional support comes from the Northwestern Polytechnical University. The authors have no conflicts of interest to declare that are relevant to the content of this chapter.

References

1. Vink, P., Bazley, C., Kamp, I., et al. (2012). Possibilities to improve the aircraft interior comfort experience. *Applied Ergonomics, 43*(2), 354–359. https://doi.org/10.1016/j.apergo.2011.07.012
2. Vink, P., Koningsveld, E. A. P., & Molenbroek, J. F. (2006). Positive outcomes of participatory ergonomics in terms of greater comfort and higher productivity. *Applied Ergonomics, 37*(4), 537–546. https://doi.org/10.1016/j.apergo.2005.10.003
3. Fazlollahtabar, H. (2010). A subjective framework for seat comfort based on a heuristic multi-criteria decision making technique and anthropometry. *Applied Ergonomics, 42*(1), 16–28. https://doi.org/10.1016/j.apergo.2010.05.004
4. Li, W., Zhang, H., Li, J., et al. (2018). An evaluation method of wearing comfort for exoskeleton robots behaviors. In *2018 IEEE international conference on robotics and biomimetics (ROBIO 2018)* (pp. 2439–2443). IEEE. https://doi.org/10.1109/ROBIO.2018.8665221
5. Kermavnar, T., O'Sullivan, K. J., de Eyto, A., et al. (2021). Relationship between interface pressures and pneumatic cuff inflation pressure at different assessment sites of the lower limb to aid soft exoskeleton design. *Human Factors, 63*(6), 1061–1075. https://doi.org/10.1177/0018720820965583
6. Lambelet, C., Temiraliuly, D., Siegenthaler, M., et al. (2020). Characterization and wearability evaluation of a fully portable wrist exoskeleton for unsupervised training after stroke. *Journal of NeuroEngineering and Rehabilitation, 17*(1), 1–16. https://doi.org/10.1186/s12984-020-00740-0
7. Crea, S., Nann, M., Trigili, E., et al. (2018). Feasibility and safety of shared EEG/EOG and vision-guided autonomous whole-arm exoskeleton control to perform activities of daily living. *Scientific Reports, 8*(1), 1–9. https://doi.org/10.1038/s41598-018-27326-5
8. Beeler, N., Roos, L., Delves, S. K., et al. (2018). The wearing comfort and acceptability of ambulatory physical activity monitoring devices in soldiers. *IISE Transactions on Occupational Ergonomics and Human Factors, 6*(1), 1–10. https://doi.org/10.1080/24725838.2018.1487580

9. Akkas, O., Lee, C. H., Hu, Y. H., et al. (2017). Measuring exertion time, duty cycle and hand activity level for industrial tasks using computer vision. *Ergonomics, 60*(12), 1730–1738. https://doi.org/10.1080/00140139.2017.1349173
10. Ball, R. M. (2011). *SizeChina: A 3D anthropometric survey of the Chinese head. [Research report]*. https://doi.org/10.13140/RG.2.1.3252.0249
11. Lacko, D. (2017). The application of 3D anthropometry for the development of headgear - A case study on the design of ergonomic brain-computer interfaces. [Research report]. https://doi.org/10.13140/RG.2.2.12196.94088
12. Wells, J. C. K., Cole, T. J., & Treleaven, P. (2008). Age-variability in body shape associated with excess weight: The UK National Sizing Survey. *Obesity, 16*(2), 435–441. https://doi.org/10.1038/oby.2007.72
13. Stewart, A., Ledingham, R., & Williams, H. (2017). Variability in body size and shape of UK offshore workers: A cluster analysis approach. *Applied Ergonomics, 58*, 265–272. https://doi.org/10.1016/j.apergo.2016.07.010
14. Fan, H., Yu, S., Chu, J., et al. (2019). Anthropometric characteristics and product categorization of Chinese auricles for ergonomic design. *International Journal of Industrial Ergonomics, 69*(1), 118–141. https://doi.org/10.1016/j.ergon.2019.01.004
15. Fan, H., Yu, S., Wang, M., et al. (2021). Analysis of the external acoustic meatus for ergonomic design: Part I - Measurement of the external acoustic meatus using casting, scanning and rapid estimation approaches. *Ergonomics, 64*(5), 640–656. https://doi.org/10.1080/00140139.2020.1869215
16. Fan, H., Yu, S., Wang, M., et al. (2021). Analysis of the external acoustic meatus for ergonomic design: Part II - Anthropometric variations of the external acoustic meatus by sex, age and side in Chinese population. *Ergonomics, 64*(5), 657–670. https://doi.org/10.1080/00140139.2020.1869216
17. Wang, M., Fan, H., Harris, C., et al. (2019). Anthropometric characteristics of Chinese auricles for ergonomic design. *Proceedings of the Human Factors and Ergonomics Society Annual Meeting, 63*(1), 1962–1963. https://doi.org/10.1177/1071181319631275
18. Fan, H., Wang, M., Adamson, C. H., et al. (2020). Anthropometric analysis of external acoustic meatus for ergonomic design. *Proceedings of the Human Factors and Ergonomics Society Annual Meeting, 64*(1), 1833–1835. https://doi.org/10.1177/1071181320641397
19. Doutres, O., Sgard, F. C., Benacchio, S., et al. (2018). Earplug comfort: From subjective assessment on the field to objective measurement and simulation using augmented artificial heads. *The Journal of the Acoustical Society of America, 143*(3), 1910–1930. https://doi.org/10.1121/1.5025311
20. Vink, P., & Hallbeck, S. (2012). Editorial: Comfort and discomfort studies demonstrate the need for a new model. *Applied Ergonomics, 43*(2), 271–276. https://doi.org/10.1016/j.apergo.2011.06.007
21. Wang, M., Fan, H., Yu, S., et al. (2022). Effects of variations in the tragus expansion angle on physical comfort for in-ear wearables. *Ergonomics*. https://doi.org/10.1080/00140139.2022.2092345
22. Schifferstein, H. N. J., & Hekkert, P. (2011). *Product experience*. Elsevier. https://doi.org/10.1016/C2009-0-62946-3
23. Clarkson, P. J., Coleman, R., Keates, S., et al. (2013). *Inclusive design: Design for the whole population*. Springer. https://doi.org/10.1007/978-1-4471-2937-0
24. Chang, H. M., Ivonin, L., Chen, W., et al. (2011). Lifelogging for hidden minds: Interacting unconsciously. *Lecture Notes in Computer Science, 6972*, 411–414. https://doi.org/10.1007/978-3-642-23953-9_51
25. Fokkinga, S., Hekkert, P., Desmet, P., et al. (2014). From product to effect towards a human-centred model of product impact. In *6th design research society conference*, 16-19 June 2014.
26. Demir, E., Desmet, P. M. A., & Hekkert, P. (2009). Appraisal patterns of emotions in human-product interaction. *International Journal of Design, 3*(2), 41–51.

27. Walls, A. R., Okumus, F., Wang, Y. R., et al. (2011). An epistemological view of consumer experiences. *International Journal of Hospitality Management, 30*(1), 10–21. https://doi.org/10.1016/j.ijhm.2010.04.001
28. Helander, M. G., & Zhang, L. (1997). Field studies of comfort and discomfort in sitting. *Ergonomics, 40*(9), 895–915. https://doi.org/10.1080/001401397188099
29. De Looze, M. P., Kuijt-Evers, L. F. M., & Van Dieën, J. (2003). Sitting comfort and discomfort and the relationships with objective measures. *Ergonomics, 46*(10), 985–997. https://doi.org/10.1080/0014013031000161371 8
30. Naddeo, A., Cappetti, N., Vallone, M., et al. (2014). New trend line of research about comfort evaluation: proposal of a framework for weighing and evaluating contributions coming from cognitive, postural and physiologic comfort perceptions. *Advances in Social and Organizational Factors, 1*, 503–515.
31. Sickel, K., Baloch, S., Melkisetoglu, R., et al. (2011). Toward automation in hearing aid design. *Computer-Aided Design, 43*(12), 1793–1802. https://doi.org/10.1016/j.cad.2011.06.009
32. Fan, H., Wang, M., Xu, S., et al. (2025). Surface pressure sensitivity of the auricle and periauricular areas for design purposes of product interaction. *Proceedings of the Human Factors and Ergonomics Society Annual Meeting, 69*(1). https://doi.org/10.1177/10711813251367739
33. Pan, C. S., Gardner, L. I., Landsittel, D. P., et al. (1999). Ergonomic exposure assessment: An application of the PATH systematic observation method to retail workers. *International Journal of Occupational and Environmental Health, 5*(2), 79–87. https://doi.org/10.1179/oeh.1999.5.2.79
34. Harris, C., & Straker, L. (2000). Survey of physical ergonomics issues associated with school childrens' use of laptop computers. *International Journal of Industrial Ergonomics, 26*(3), 337–346. https://doi.org/10.1016/S0169-8141(00)00009-3
35. Harris-Adamson, C., Eisen, E. A., Kapellusch, J., et al. (2015). Biomechanical risk factors for carpal tunnel syndrome: A pooled study of 2474 workers. *Occupational and Environmental Medicine, 72*(1), 33–41. https://doi.org/10.1136/oemed-2014-102378
36. Wang, M., Zhao, C., Barr, A., et al. (2021). Hand posture and force estimation using surface electromyography and an Artificial Neural Network. *Human Factors, 63*(6), 1014–1034. https://doi.org/10.1177/0018720820905250
37. Li, W., Yu, S., Yang, H., et al. (2017). Effects of long-duration sitting with limited space on discomfort, body flexibility, and surface pressure. *International Journal of Industrial Ergonomics, 58*, 12–24. https://doi.org/10.1016/j.ergon.2017.01.008
38. Zhao, C., Yu, S., Li, W., et al. (2018). Predicting aircraft seat comfort using an artificial neural network. *Human Factors and Ergonomics in Manufacturing & Service Industries, 28*(6), 275–283. https://doi.org/10.1002/hfm.20746
39. Alexander, M., & Laubach, L. L. (1968). *Anthropometry of the human ear (a photogrammetric study of USAF flight personnel). AMRL-TR-68-1*. Aerospace Medical Research Laboratory.
40. Ferrario, V. F., Sforza, C., Ciusa, V., et al. (1999). Morphometry of the normal human ear: a cross-sectional study from adolescence to mid-adulthood. *Journal of Craniofacial Genetics and Developmental Biology, 19*(4), 226–233.
41. Wang, M., Fan, H., Yu, S., et al. (2022). Analysis of the auricles and auricular shape types for ear-related wearables: A study of mainland Chinese samples aged 15-79. *Work, 73*(1), 1–20. https://doi.org/10.3233/WOR-205241
42. Purkait, R., & Singh, P. (2007). Anthropometry of the normal human auricle: A study of adult Indian men. *Aesthetic Plastic Surgery, 31*(4), 372–379. https://doi.org/10.1007/s00266-006-0231-4
43. Wang, B., Dong, Y., Zhao, Y., et al. (2011). Computed tomography measurement of the auricle in Han population of north China. *Journal of Plastic, Reconstructive & Aesthetic Surgery, 64*(1), 34–40. https://doi.org/10.1016/j.bjps.2010.03.053
44. Alexander, K. S., Stott, D. J., Sivakumar, B., et al. (2011). A morphometric study of the human ear. *Journal of Plastic, Reconstructive & Aesthetic Surgery, 64*(1), 41–47. https://doi.org/10.1016/j.bjps.2010.04.005

45. Lee, W., Jung, H., Bok, I., et al. (2016). Measurement and application of 3D ear images for earphone design. *Proceedings of the Human Factors and Ergonomics Society Annual Meeting, 60*(1), 1053–1057. https://doi.org/10.1177/1541931213601243
46. Purkait, R. (2013). Progression of growth in the external ear from birth to maturity: A 2-year follow-up study in India. *Aesthetic Plastic Surgery, 37*(3), 605–616. https://doi.org/10.1007/s00266-013-0097-1
47. Standring, S. (2016). *Gray's anatomy: The anatomical basis of clinical practice* (41st ed.). Elsevier.
48. Zhao, S., Li, D., Liu, Z., et al. (2018). Anthropometric growth study of the ear in a Chinese population. *Journal of Plastic, Reconstructive & Aesthetic Surgery, 71*(4), 518–523. https://doi.org/10.1016/j.bjps.2017.10.010
49. Ahmed, A. A., & Omer, N. (2015). Estimation of sex from the anthropometric ear measurements of a Sudanese population. *Legal Medicine, 17*(5), 313–319. https://doi.org/10.1016/j.legalmed.2015.03.002
50. Lee, W., Yang, X., Jung, H., et al. (2018). Anthropometric analysis of 3D ear scans of Koreans and Caucasians for ear product design. *Ergonomics, 61*(11), 1480–1495. https://doi.org/10.1080/00140139.2018.1493150

Chapter 2
Models of Physical Comfort Design for Ear-Related Wearables Integrated with Ear Anthropometry

Abstract A systematic framework is proposed to model physical comfort in ear-related wearable devices. As ear-worn products evolve from single-purpose audio transmission to multifunctional interaction and biosignal monitoring, physical comfort has become a critical determinant of user acceptance and long-term wear compliance. The external ear presents distinctive ergonomic challenges due to its limited contact area, complex geometry, heterogeneous tissue composition, and high sensory sensitivity, rendering conventional general ergonomics-based comfort models insufficient. This chapter introduces static and dynamic comfort perception models that attribute discomfort to mismatches between product parameters and the tolerance thresholds of ergonomic design factors (EDFs). Dynamic comfort is further shaped by usage-related variables, including wearing motivation, duration, and environmental context. A human-product interaction path model is developed to illustrate how anatomical variability, such as bilateral asymmetry, sex differences, and age-related morphological changes, constrains one-size-fits-all design approaches. Key EDFs are extracted through semi-structured interviews and structured questionnaires, and factor analysis is employed to categorize them into three principal dimensions: auricle shape-structure, external acoustic meatus (EAM) shape-structure, and transitional anatomical structures. The proposed framework provides theoretical and methodological foundations for predicting physical comfort, defining tissue deformation thresholds, and guiding the ergonomic design of inclusive ear-related wearable devices.

Keywords Physical comfort · Ear-related wearables · Ergonomic design factors (EDFs) · Ear anthropometry · Comfort perception modelling

In recent years, ear-related wearable devices have rapidly expanded beyond their traditional functions of sound transmission and hearing protection. Modern products now encompass diverse applications, ranging from sports and entertainment to voice interaction, communication, and bio-signal monitoring. This evolution has transformed the external ear from a peripheral auditory organ into a multifunctional interface for human-product interaction. With this transformation, the concept of

H. Fan et al., *Engineering Anthropometry of the External Ear for Human-Product Interaction*, https://doi.org/10.1007/978-981-92-2160-8_2

physical comfort has emerged as a pivotal determinant of both product success and user acceptance.

Physical comfort is more than a matter of preference; it represents a fundamental human requirement in product ergonomics. Devices that compromise comfort are unlikely to achieve long-term adherence, regardless of their technical sophistication or functional innovation. The challenge is particularly acute for ear-related wearables because of the unique structural and physiological features of the external ear. Unlike larger anatomical regions, the external ear provides a limited surface area, irregular geometry, and heterogeneous tissue composition, including cartilage, skin, and muscle. Furthermore, its high sensory acuity, due to dense and complex innervation, magnifies discomfort even under relatively minor pressures or deformations. As a result, the external ear demands a dedicated framework for evaluating and modelling physical comfort.

Conventional ergonomics has long addressed comfort in areas such as seating, workstations, and protective equipment. These domains have produced theoretical and quantitative models that link anthropometric dimensions, pressure distribution, and environmental factors to comfort perception. However, direct application of such models to the external ear is problematic. The auricle and external acoustic meatus differ from musculoskeletal components in that they do not primarily serve biomechanical or load-bearing functions. Instead, they act as delicate sensory and structural regions with limited tolerance for deformation. Existing ergonomic approaches, therefore, fail to fully capture the intricate interplay between product design parameters and the anthropometric and physiological variability of the external ear.

This gap becomes even more significant when considering population diversity. Ear morphology is subject to bilateral asymmetry, sex-based differences, and age-related growth trajectories. Studies have demonstrated that auricular dimensions mature at varying ages, with length, width, and conchal size reaching maturity at different stages of development. Moreover, auricular growth continues throughout life, albeit at a slower rate after adolescence. Such variability implies that one-size-fits-all design strategies are inherently flawed. Without population-specific anthropometric data and models, the likelihood of product-user mismatch remains high, leading to discomfort, reduced usability, and potential health risks.

Another challenge lies in the dynamic nature of product use. Ear-related wearables are not static accessories; they are often employed during physical activity, in varying environments, and for prolonged durations. Dynamic usage contexts, such as jogging, working in noisy environments, or extended communication, exert cumulative pressures on the ear and significantly influence comfort perception. While a device may appear tolerable under short-term static conditions, the same product may induce discomfort, fatigue, or even pain under dynamic conditions. The lack of integrated models that account for both static and dynamic dimensions of comfort represent a critical shortcoming in current research and design practice.

Given these challenges, there is a pressing need for a systematic framework that unites anthropometric precision with ergonomic evaluation. Such a framework must move beyond descriptive accounts of discomfort toward quantitative models capable of predicting comfort outcomes under diverse conditions. Central to this approach is the concept of ergonomic design factors (EDFs), defined as measurable attributes of the human user that influence comfort perception. By identifying, classifying, and quantifying EDFs, researchers and designers can establish scientifically grounded thresholds and tolerance intervals for product parameters. This approach enables the transformation of comfort from a subjective, variable experience into a quantifiable and designable dimension of product development.

The integration of EDFs with ear anthropometry offers a promising pathway toward this goal. Ear anthropometry provides the necessary dimensional data to characterize auricular and ear canal variability across populations, while EDF-based modelling captures the ergonomic interactions that give rise to comfort or discomfort. Together, these perspectives create a dual foundation for advancing both theoretical understanding and practical design. Such integration ensures that comfort assessment is not merely a matter of post-hoc user feedback, but a proactive element embedded in the design process.

The study of physical comfort in ear-related wearables must be reframed as a multidisciplinary endeavour that combines ergonomics, anthropometry, physiology, and design science. The external ear presents unique challenges that render conventional comfort models insufficient. At the same time, it provides an opportunity to develop new frameworks that not only address immediate design needs but also contribute to broader theories of human-product interaction. By situating comfort at the intersection of measurable human attributes and product design parameters, this chapter sets the stage for a systematic exploration of models that will inform subsequent analysis, measurement, and application in the design of ear-related wearables. Collectively, this chapter serves as both a theoretical foundation and methodological framework for subsequent chapters. It bridges the conceptual understanding of comfort with measurable anthropometric and ergonomic data, enabling the systematic development of design guidelines for ear-related wearables. By embedding ear anthropometry into the modelling process, it ensures that comfort evaluation is not an abstract notion but a quantifiable and reproducible dimension of product ergonomics. Ultimately, the models and empirical findings presented herein provide researchers and designers with robust tools for balancing functional innovation with the fundamental human requirement of physical comfort.

2.1 Models of Ergonomic Design Factors for Wearing Comfort

Wearing comfort of the external ear belongs to the broader domain of physical comfort. It is defined as the subjective feedback of (dis)comfort experienced by the user during instrumental activities (wearing) in the process of human-product interaction.

2.1.1 Perception of Wearing Comfort

The perception of physical comfort originates from numerical changes in EDFs during product use. The degree of influence on overall physical comfort differs across factors [1–3]. For the external ear wearing comfort, such changes arise from mismatches between product parameters and EDFs [4–5]. Thus, static perception of wearing comfort under simultaneous exposure to multiple EDFs can be represented by a static multi-factor model:

$$P_{Fitting} = \sum_{i=1}^{m} W_i P_i \tag{2.1}$$

where, P denotes the user's perceived wearing (dis)comfort; i indicates a specific EDF of the user; W represents the weight of a specific EDF; m is the total number of EDFs.

Static perception of wearing comfort under a single EDF can be described by a *static single-factor model*:

$$P_i = f_i\left(V_i | R_i\right) \tag{2.2}$$

where, V represents the difference between product parameters and the EDFs; R represents the comfort tolerance interval of the user's ergonomic factor changes (comfort threshold).

According to Demir et al. and Naddeo et al. dynamic wearing comfort perception is typically lower than static perception due to motivational factors during product use (e.g., 30 min of jogging while wearing the product) [6–8]. Hence, dynamic wearing comfort perception of the external ear should also account for losses induced by specific user motivations (usage time and context). Dynamic perception under a specific usage motive can be represented by a *dynamic mechanism model*:

$$P_{Motive} = P_{Fitting} + L_j\left(T_j, SC_j\right) \tag{2.3}$$

where, j represents a specific usage motivation; T represents the usage duration; SC represents the usage scenario/context; $L_j(T_j, SC_j)$ is the comfort loss induced by the specific usage motive, where $L_j(T_j, SC_j) \leq 0$.

2.1.2 Human-Product Interaction Path

User comfort experience with a product is not only the outcome of interaction but also a determinant of subsequent usage choices and product design. According to the Schifferstein-Hekkert theory [9], the interaction path of the external ear wearing comfort can also be described as a dual-cue structure (Fig. 2.1). The frontend of the

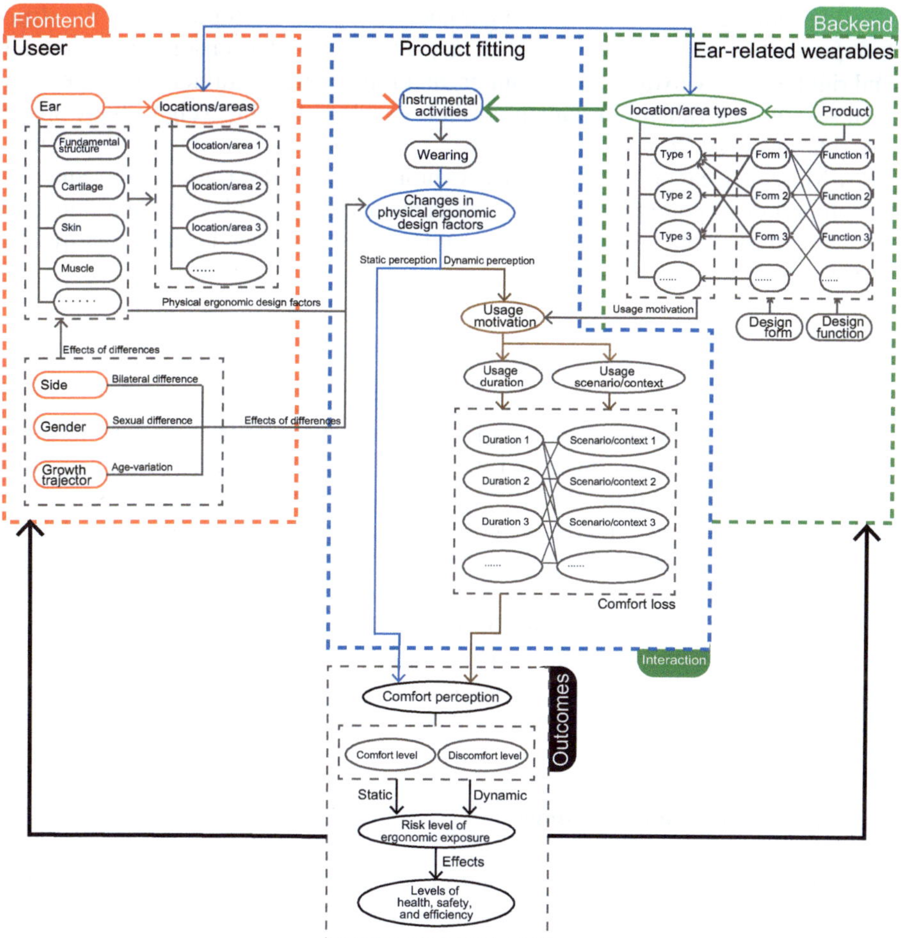

Fig. 2.1 Path model of ear-related wearables based on human-product interaction

dual-cue structure consists of the users and their perception of information, while the backend consists of the ear-related wearables and associated information.

Frontend

The external ear is composed of multiple structural elements such as cartilaginous framework, cartilage, skin, and muscles. These elements influence or constrain both user comfort and product design. These anatomical features define potential fitting locations/areas for the human-product interaction, such as the auricular contour, concha, and external acoustic meatus entrance.

Biomechanically, human system parameters are not isotropic, nor is the structure perfectly symmetrical [10, 11]. As a paired peripheral sensory organ, the left and right ears may exhibit significant or minor differences. Numerous studies confirm sexual dimorphism in human attributes such as height, weight, body shape, skin elasticity, and muscle activity [10–13], which may also extend to ear structures. Moreover, ear growth trajectories exhibit age-related variation. Consequently, the properties of potential Fitting locations/areas for wearables are determined by ear structures and influenced by bilateral, sex-related, and age-related differences.

Backend

Ear-related wearables offer users various product types defined by functions and forms. Depending on functional and structural combinations, ear-related wearables can be categorized by fitting location/area [14–16]. For instance, in-ear sports earphones and in-ear voice interaction devices provide functions such as sound transmission, noise control, and bio-signal collection, primarily fitting the cavum concha and external acoustic meatus (EAM, ear canal). Earplugs provide noise control and hearing protection, mainly fitting the EAM. Different fitting types pose distinct demands on ear anatomy and also serve as reference points for ergonomic research. Thus, beyond direct interaction, fitting locations/areas and product types mutually influence both fundamental research and design applications.

Interaction Mode and Outcomes

The interaction mode for physical comfort perception of the external ear is product fitting. Product fitting triggers the wearing activity, causing changes in ear attributes and resulting in variations in physical ergonomic factors. According to Eq. (2.2), when physical ergonomic factor variations fall within the tolerance range, users perceive comfort or discomfort; when variations exceed thresholds, mismatches may prevent usage or cause injury, precluding comfort perception.

Static comfort exposure results from anatomical changes induced by fitting. Dynamic comfort states additionally require consideration of motivational losses (Eq. (2.3)). Comfort loss often relates to fatigue, usually mediated by multiple ergonomic factors. Usage scenarios/contexts are commonly classified into low-demand and high-demand types. The classification criteria for types of physical demands are generally based on whether the usage scenario/context depends on environmental ergonomic factors [17]. For instance, industrial workplaces (high-demand) involve dominant discomfort inducers such as noise, thermal comfort, skin moisture, vibration, and lighting [18–21]. In these cases, total discomfort arises from combined ergonomic and environmental factors. In low-demand scenarios/contexts, discomfort is mainly induced by ergonomic factors, followed by cognitive factors [22–23], such as familiarity with tasks and urgency. When participants are relaxed and familiar with tasks, self-report data can be used to analyse comfort loss [24, 25].

2.1.3 Extraction Path and Structure

Extraction Path

EDFs required for comfort and exposure are derived from user attributes. Two common extraction approaches are used, both adopting a frontend (user) perspective:

1. Direct extraction is typically applied to studies involving a single attribute category, e.g., RULA-based upper limb posture analysis, where joint angles are directly selected as EDFs.
2. Weighting/statistical-relationship construction is used for multiple attribute categories, e.g., cabin seating comfort, involving anthropometric dimensions, pressure tolerance, and joint angles. Here, subjective data are analysed to identify comfort-related factors, which are then converted into variables with calculated weights.

For ear-related wearables, physical comfort issues mainly arise from dimensional changes and pressure effects, leading to deformations of ear structures. Thus, relevant attribute categories are ear anthropometric dimensions and skin surface-pressure sensitivity. This book adopts the weight/statistical relation approach to extracting EDFs, using a frontend perspective to address general-purpose design.

The proposed path is as follows:

1. Determine attribute categories;
2. Identify influencing factors (fitting locations/areas with comfort issues) via factor analysis;
3. Define descriptive variables for measurement;
4. Filter variables by contributions/loadings;
5. Group variables through correlation analysis;
6. Construct grouped factors for the perception model;
7. Consider bilateral, sex-related, and age-related differences.

Structure

As shown in Sect. 2.1.2, the structure of EDFs could be described as a *nested serial structure* (Fig. 2.2). All EDFs are in series, and sub-factors within each factor are also serially arranged. If any factor exceeds its threshold, the user's comfort perception fails.

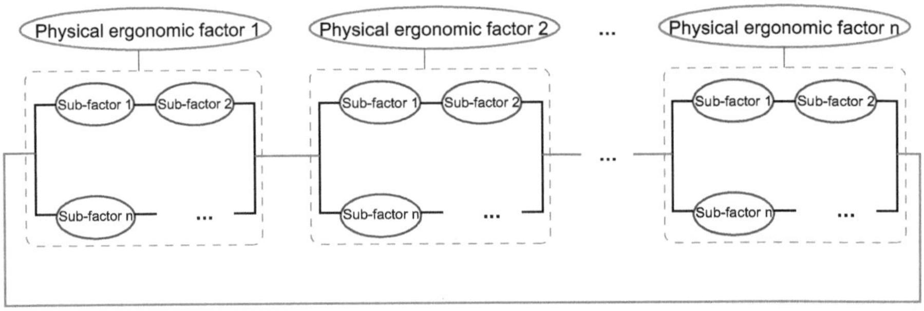

Fig. 2.2 Structure model of EDFs of wearing comfort for human external ear

The subfactor-based static perception can be expressed as:

$$P_i = \sum_{k=1}^{n} \omega_{ik} p_{ik} \tag{2.4}$$

where, P presents the user's perceived wearing (dis)comfort; p presents the perception associated with sub-factor variation; i presents an EDF; k presents a sub-factor; n is the number of sub-factors; ω is the weight of sub-factor.

2.2 Structures of the External Ear Related to Wearing Comfort

In research on human wearable devices, the composition of target fitting locations/areas (e.g., head, hand) is central to comfort studies. The basic structures of the target fitting region, along with the skin, cutaneous innervation, and muscle distribution, influence design forms, product structures, and deformation-related design parameters. The developmental maturity period or mechanism of the target fitting region constrains the age-group classification of users and guides functional design strategies.

Unlike organs of the human musculoskeletal system (e.g., upper limbs, neck), the external ear does not function as a biomechanical component. Therefore, its structural themes cannot be directly classified following biomechanical or kinesiology models. For example, in the upper limb, comfort-related structural themes include movement modules (shoulder joint, distal segments), skeletal structures (arm, elbow, forearm, wrist, hand), joint types (fibrous, synovial), skin and cutaneous innervation, muscle groups (superficial, deep), axes and orientations.

2.2.1 Basic Structures Related to Wearing Comfort

Fundamental Structure

According to conventional anatomical and engineering anthropometric classification, the human ear consists of the external ear, middle ear, and inner ear [10, 26]. Except for the auricle (or pinna), most parts of the human ear are embedded within the temporal bone. As paired organs located bilaterally on the mid-sagittal plane, ears receive, modulate, transmit, amplify, and differentiate complex sound waves, and also serve as terminal sensory organs for distance perception and balance. The external ear consists of the auricle and the external acoustic meatus (EAM, or ear canal).

The auricle's distinctive position and physiological function make it the primary fitting site for wearable devices [13, 27, 28]. Projecting laterally from the temporal bone, the auricle's prominence varies across individuals and may significantly differ across genders, ages, and ethnicities. The EAM collects sound and directs it to the tympanic membrane (eardrum). Hence, the auricle is considered the first regulator of auditory response to external acoustic stimuli.

Morphologically, the auricle has a cartilaginous, shell-like, spiral structure. Components of the auricle include the helix, crus of helix, tubercle (Darwin's tubercle), antihelix, crura of antihelix, triangular fossa, scaphoid fossa, concha, tragus, antitragus, and lobule (Fig. 2.3). The concha is divided by the crus of helix into the cavum concha and cymba concha. The anterior notch (incisura anterior auris) lies between the helix and tragus cartilages; the intertragic notch lies between the tragus and antitragus. The otobasion is the auricular base connecting the cartilage to the temporal bone. The lobule, made of fibrous and adipose tissue, is soft but inelastic, while other parts of the auricle consist of cartilage and overlying skin, providing elasticity and resilience.

Unlike the auricle, the EAM lies within the temporal bone and serves as the main sound conductor to the tympanic membrane. In wearables, the EAM anchors functional units such as speakers (earphones), noise-cancelling elements (earplugs), and

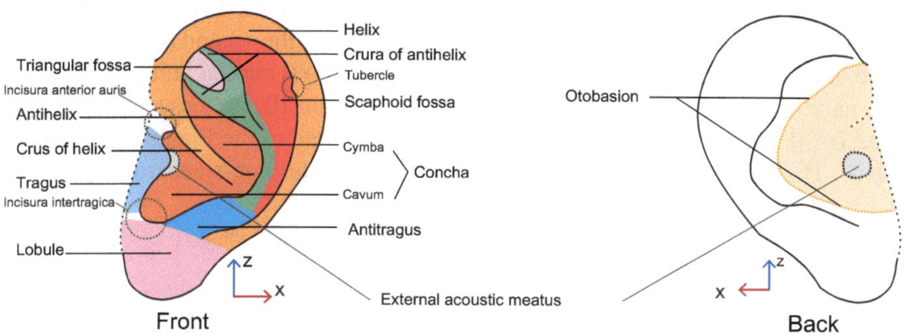

Fig. 2.3 Structure of the auricle

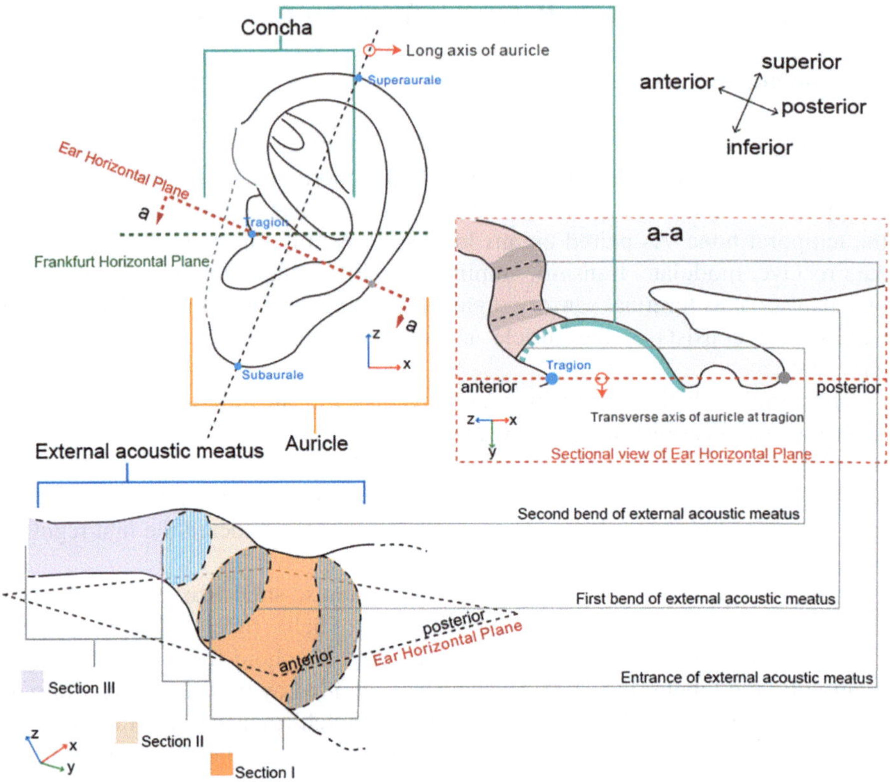

Fig. 2.4 Structure of the external acoustic meatus

sound-conduction components (hearing aid tubes). The EAM extends from the concha to the eardrum, forming an "S-shaped curve". CT data of Caucasians indicate a distance of approximately 4 cm from the tympanic membrane to the tragus base, and approximately 2.5 cm to the posterior concha [10].

The EAM includes entrance, three sections, and two bends (Fig. 2.4) [29, 30]. The EAM Section I spans from the entrance to first bend; the EAM Section II from the first bend to second bend; the EAM Section III from the second bend to eardrum. The outer one-third (Sections I & II, partly Section III) is cartilaginous, flexible like the auricle, while the inner two-thirds (mainly Section III) are osseous, rigid [10]. The EAM Section I aligns parallel or at an acute angle to the Frankfurt horizontal plane. The cross-section of EAM is elliptical or irregular/non-standard elliptical.

Cartilaginous Framework

The external ear cartilage framework includes auricular cartilage and the EAM cartilage. Auricular cartilage is a thin elastic plate covered by skin, connected to adjacent tissues by ligaments and muscles. Visible auricular depressions and protrusions

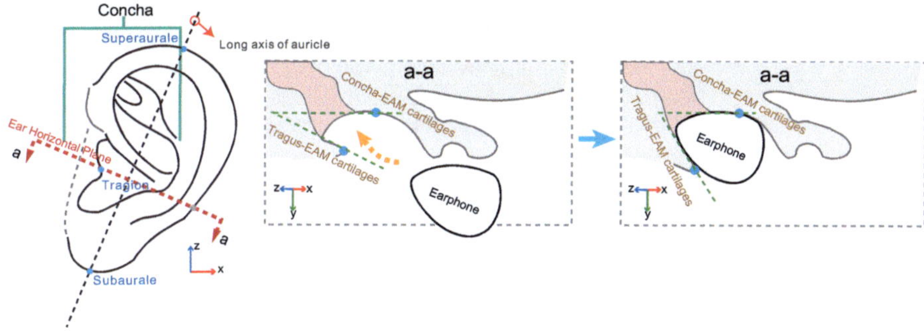

Fig. 2.5 Expansion of tragus-concha-EAM cartilages caused by ear-related wearables fitting

arise from the cartilaginous form. The lobule and region between the tragus and the crus of helix lack cartilage, being filled with dense fibrous tissue.

For in-ear wearable devices that are inserted into the EAM (e.g., earphones and in-ear hearing aids), two substructures within the auricular cartilage framework (serving as connecting structures) play key roles in fixation and anti-drop functions (Fig. 2.5): the Tragus-EAM cartilage structure and the Concha-EAM cartilage structure [5, 10]. The Tragus-EAM cartilage structure consists of the tragus plate and the anterior wall of the EAM cartilage, located on the anterior side of the external ear. The Concha-EAM cartilage structure consists of the conchal portion of the auricular cartilage and the posterior wall of the EAM cartilage, located on the posterior side of the external ear. These two substructures form an angle, and through elastic deformation of the auricular cartilage, the outward expansion of this angle provides fixation for the device that needs to be fitted.

In the human skeletal system, there are three types of cartilage tissues: hyaline cartilage, elastic cartilage, and fibrocartilage. Elastic cartilage is composed of chondrocytes and extracellular matrices, characterized by a high content of elastic fibres, thereby conferring high elasticity. The auricular cartilage belongs to elastic cartilage. Since it is not part of the musculoskeletal components responsible for axial rotation, lateral bending, stretching, or contraction, its biomechanical properties remain unclear. In gold-standard studies of auricular cartilage reconstruction, costal cartilage is generally used for morphological reconstruction, while biomechanical property simulation or estimation is rarely addressed [29]. Consequently, reconstructed or remodelled auricles may lack sufficient elasticity in the cartilage framework. This limitation is particularly relevant when using products involving the tragus, concha, and external acoustic meatus, where the fit of ear-worn wearables may be restricted to varying degrees [30].

Skin and Cutaneous Innervation

Auricular skin extends from the auricle to the EAM, covering the tympanic membrane. It is thin, thicker in the cartilaginous EAM, with weak thermoregulation due to the absence of dermal papillae, and exhibits high pain sensitivity [10].

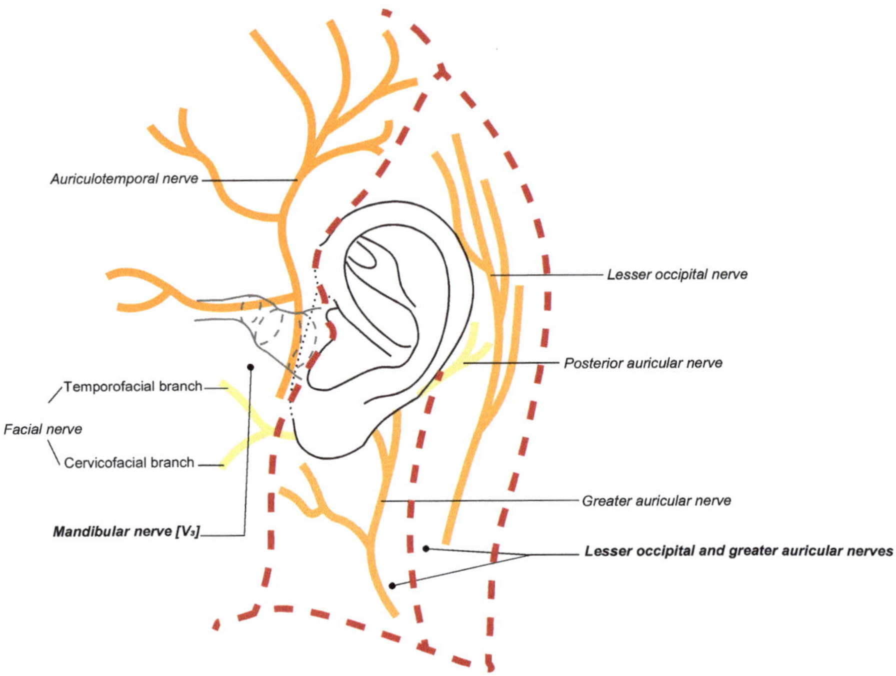

Fig. 2.6 Cutaneous innervation (in bold and italic) of the external ear

Cutaneous innervation of the auricle is complex and not fully established. Broadly, it involves branches of the mandibular nerve [V_3] and the lesser occipital-greater auricular nerves (Fig. 2.6). The greater auricular and lesser occipital nerves innervate most of the auricle, excluding the tragus. The auriculotemporal branch of [V_3] innervates the tragus and parts of the EAM. Additional contributions may arise from the temporofacial and cervicofacial branches of the facial nerve, posterior auricular nerve, and auricular branch of the vagus nerve. Thus, auricular skin sensation is a convergence of multiple nerve inputs, and its sensory model must account for multiple innervation pathways.

Muscles

Auricular muscles are classified as extrinsic and intrinsic, distributed on and around the auricle [31–33] (Fig. 2.7). Extrinsic muscles that include auricularis superior, auricularis anterior, and auricularis posterior, enable gross movement of the auricle with facial motion. Intrinsic muscles connect auricular subregions but do not significantly affect shape, including helicis major, helicis minor, tragicus (anterior), antitragicus, and obliquus auriculae (posterior). Auricular sEMG can detect cross-ear responses, though extrinsic muscles are difficult to measure noninvasively.

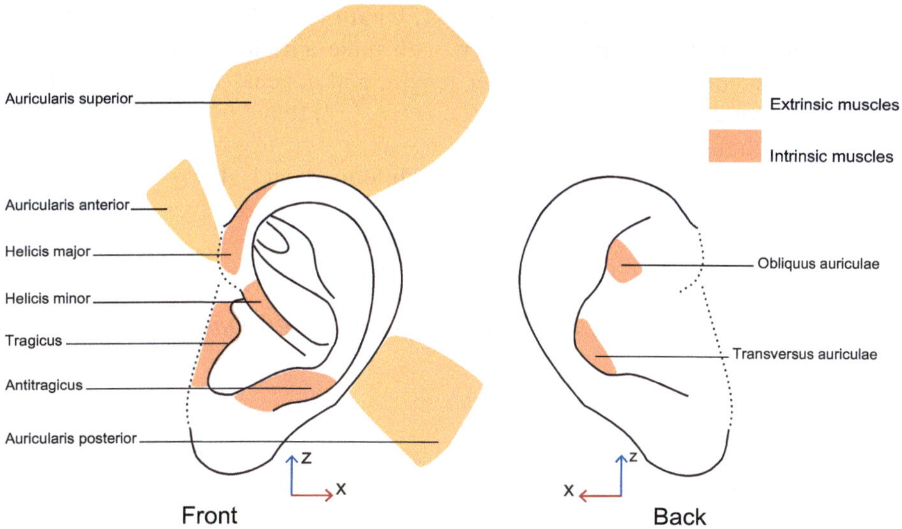

Fig. 2.7 Muscles of the external ear

Notably, auricularis posterior muscle is sensitive to external stimuli such as pressure, vibration, and sound [34, 35].

2.2.2 *Growth Trajectories*

Maturation of the external ear structure is identified when the mean value of a linear variable (e.g., auricular length, width, conchal dimensions) at a given age matches the corresponding mean value at age 18 within the Farkas SEM criterion [36]:

$$\overline{x}_{(i,a)} \times 2 \times SEM_{(i,a)} \approx \frac{\overline{x}_{(i,18)}}{2 \times SEM_{(i,18)}} \tag{2.5}$$

$$\overline{x}_{Mature} = \min_{a=0}^{18} \left\{ \overline{x}_{(i,a)} \right\} \tag{2.6}$$

where, $\overline{x}_{(i,a)}$ denotes the mean of a linear ear anthropometric variable i at age a; $\overline{x}_{(i,18)}$ presents the mean of the same variable at age 18; $SEM_{(i,18)}$ presents the standard error of the mean. If no significant differences exist between $\overline{x}_{(i,a)}$ and $\overline{x}_{(i,18)}$ ($p < 0.05$), the minimum value of $\overline{x}_{(i,a)}$ is selected as the maturation size ($\overline{x}_{Mature}$); otherwise, $\overline{x}_{(i,18)}$ is taken as the maturation size.

At birth, auricular linear dimensions are 60%–75% of their maturation values [96, 107]. Maturation age varies across structures, typically between ages 5–16

[10]. Width variables mature by ages 5–11, length variables by ages 12–16 [30]. Length grows faster but matures later, with male auricles maturing later than females. The auricular length, lobular length, and conchal length mature latest (16 years). For Chinese populations, Zhao et al. [37] found overall maturation by 14 years, with the auricular length maturing latest. The auricle continues lifelong growth, with rapid development during childhood and slower rates thereafter [14].

2.2.3 *Fitting Range and Comfort-Related Deformation*

Structure

Based on current technological conditions and research limitations, the fitting structure of ear-related wearables generally extends from the auricle to the second bend of the EAM [26, 29, 30]. Conventional head-mounted designs (e.g., headsets, bone-conduction earphones, behind-the-ear hearing aids) primarily rely on the auricle as the fitting structure. Devices such as (in-ear) earphones, earplugs, and in-ear hearing aids typically fit around the concha and tragus of the auricle, as well as the lateral one-third of the EAM (including Section I, Section II, and part of Section III). For noise reduction and hearing protection devices, general protective designs must at least fit to the first bend of the EAM, while deep-protection devices must extend to the second bend of the EAM.

Fitting Locations/Areas and Product Classification

Ear-related wearables have evolved from simple product components providing sound transmission or noise isolation into independent interactive tools with multi-functional capabilities. As functional demands and combinations diversify, the types of ear-related wearables have also expanded, including in-ear earphones, headsets, earplugs, bone-conduction earphones, and behind-ear hearing aids.

According to users' wearing styles, design forms can be classified into in-ear and head-mounted. In this conventional classification, in-ear products fit the auricle (tragus, concha) and the EAM, while head-mounted products fit the auricle and the head. However, with advancements in wireless transmission technologies (e.g., Bluetooth) and continuous innovation in structural design, ear-related wearables are gradually reducing their reliance on head-mounted fitting. This trend has led to a new classification system based on product form and fitting locations/areas: ear-cup (cup type), ear-muff (muff type), ear-hook/ear-band (band type), ear-bud/ear-plug (plug type). In this scheme (Fig. 2.8): (1) cup-type products fit the auricle; (2) muff-type products fit the auricle, with some also contacting the head (e.g., headsets); (3)

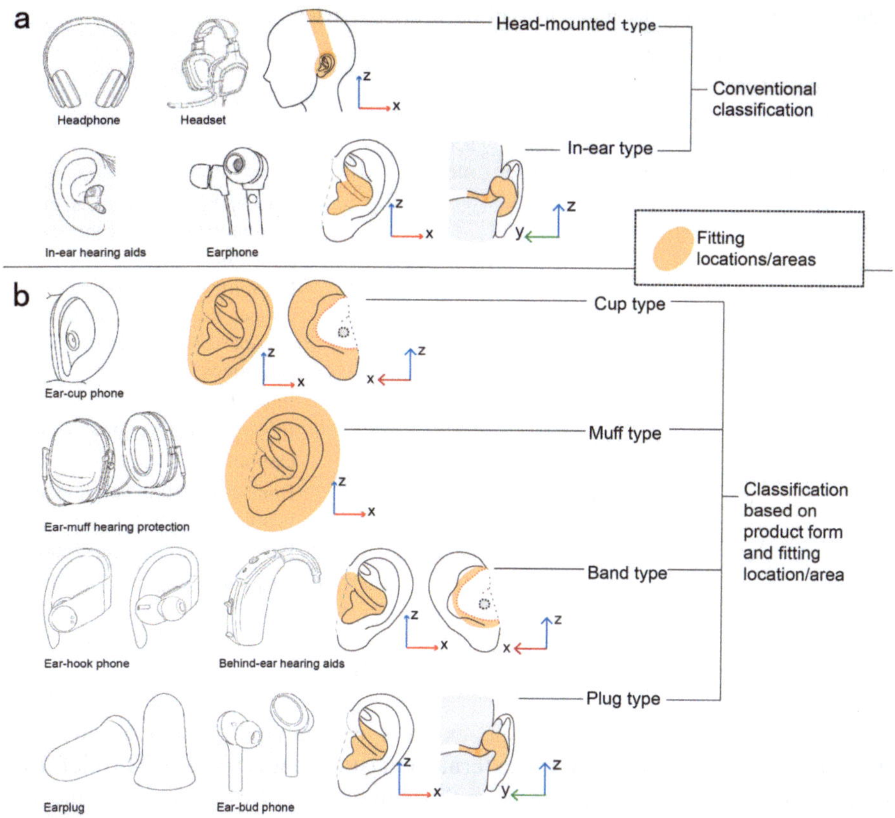

Fig. 2.8 Classification and fitting locations/areas of ear-related wearables: (**a**) conventional classification; (**b**) classification based on product fitting location/area and structure characteristics

band-type products fit the otobasion, concha, and tragus (e.g., behind-the-ear hearing aids, bone-conduction earphones), with some also extending into the EAM (e.g., sports in-ear earphones); (4) plug-type products fit the concha, tragus, and EAM (e.g., in-ear earphones), with some restricted entirely to the EAM (e.g., noise-isolating earplugs, canal earphones, or hearing aids).

According to Doutres et al. [1] and Wang et al. [4], cup-type and muff-type devices are classified as non-intrusive, whereas band-type and plug-type devices are considered intrusive. The classification criterion is whether the measurement of comfort perception requires the device to intrude (embed) into the inner auricle or the EAM. As summarized in Table 2.1, the fitting locations/areas of ear-related wearables can be categorized into the following seven types: (1) entire auricle; (2) auricle and head; (3) tragus and concha; (4) tragus, concha, and EAM; (5) a otobasion, tragus, and concha; (6) otobasion, tragus, concha, and EAM; (7) EAM only. Given the independent fitting trend of ear-related wearables, subsequent sections of this book will focus on fitting locations/areas excluding the head.

Table 2.1 Types of fitting locations/areas of ear-related wearables

No.	Fitting locations/areas						Product type
	External ear					Head	
	Auricle				External acoustic meatus		
	Entire auricle	Concha	Tragus	Otobasion			
1	✓						Ear-cup headphones, ear-muff headphones and hearing protectors
2	✓					✓	Ear-muff headphones, hearing protectors, over-ear headphones, over-ear headsets, in-ear earphones
3		✓	✓				In-ear hearing aids
4		✓	✓		✓		In-ear earphones, plug-type earphones
5		✓	✓	✓			Bone-conduction earphones, behind-ear hearing aids
6		✓	✓	✓	✓		Ear-hook/ear-band headphones
7					✓		Earplugs, in-ear hearing aids

Comfort-Related Deformations within the Fitting Structures

Fitting ear-related wearables to the external ear inevitably induces structural deformation of the auricle and the external acoustic meatus (EAM). This deformation is manifested as an increase in the values of corresponding engineering anthropometric variables. Such structural deformations alter users' perception of physical comfort, especially for devices fitted to the concha and the EAM [4]. Therefore, determining the acceptable deformation values (thresholds) of the external ear (i.e., the post-deformation values of engineering anthropometric variables) is of critical importance for improving product physical comfort. At present, three reference values are commonly adopted for the comfort range of linear deformation rates. No discomfort: 0% (entire auricle); slight discomfort: 7% (concha and EAM); complete discomfort: 10% (concha and EAM) [38]. For angular deformation rates, no reference values are usually specified angular variables are typically regarded as functional variables (e.g., defining orientation or elevation). Slight discomfort deformation rates are calculated similarly to systematic errors in repeated measurements, using a constructed coefficient of variation (CV) method. By constructing the coefficient of variation between product dimensions and subjects' anthropometric measurements, the deformation value and rate can be determined.

2.2.4 *Orientation of the External Ear*

Anatomical and engineering anthropometric orientations of human organs are mainly used for describing intrinsic structural features and for determining fitting positions of wearable devices. These orientations include superior, inferior, anterior, posterior, medial, and lateral. The orientation of the head and face is conventionally defined using the Frankfurt horizontal plane and the mid-sagittal plane as reference planes. For a long time, the external ear was described using this head-face system orientation, where the projection of the auricular length onto the midsagittal plane is perpendicular to the Frankfurt horizontal plane. However, Alexander et al. [39] noted that applying the head-face orientation system to auricular width measurements can result in underestimations or overestimations, specifically, the posterior endpoint of auricular width may not coincide with the postaurale. Consequently, subsequent researchers adopted an auricular-axis construction method to define auricular orientation (Fig. 2.9):

1. Construct the long axis of the auricle by connecting the superaurale and subaurale;
2. Construct the transverse axis of the auricle at the tragion as a line perpendicular to the long axis through the tragion;
3. Define the superaurale as superior, and the tragion as anterior.

The angle between the long axis of the auricle and the head-face coronal plane is defined as the auricular inclination angle. This orientation method intuitively reflects the directional characteristics of the external ear itself and its relative spatial relationship with the head-face region. Moreover, it facilitates effective fitting between wearable devices and auricular structures.

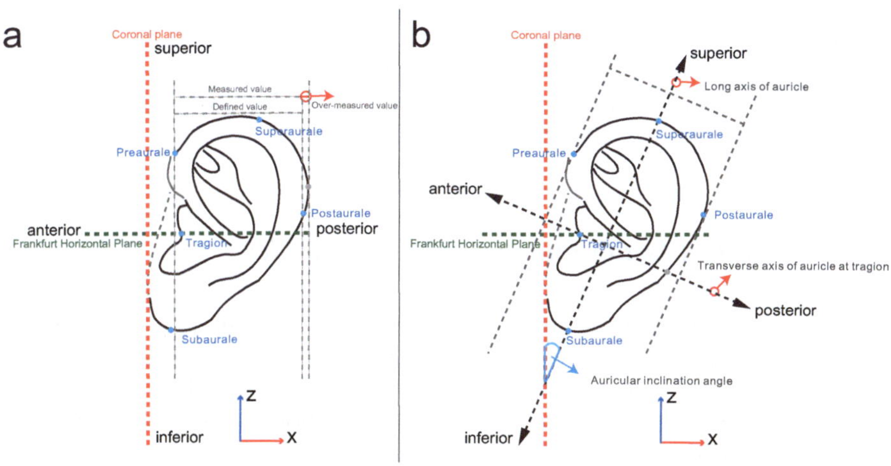

Fig. 2.9 Position and direction of the external ear in anatomy and engineering anthropometry: (**a**) orientation based on the head-face system; (**b**) orientation based on auricular axes

2.3 Extraction of Ergonomic Design Factors (EDFs) for Wearing Comfort

According to the extraction pathway outlined in Sect. 2.1, the identification of ergonomic design factors (EDFs) depends on recognizing the fitting locations of ear-related wearables where comfort issues typically arise. In practice, the process of extraction could be organized into two complementary parts that combine qualitative exploration with quantitative validation. The first part makes use of semi-structured interviews to explore experiences of the external ear comfort. An interview outline provides a consistent structure, while users of ear-related products are invited to reflect on their daily use. The discussion covers product design types, fitting locations, daily usage duration, contexts and scenarios of use, motivations for wearing, and the physical sensations of comfort or discomfort. From these discussions, recurrent comfort issues emerge, and descriptive variables suitable for measurement are defined. The second part applies a structured questionnaire survey to verify and generalize the variables identified in the interviews. Based on the earlier findings, the questionnaire is designed with targeted items that represent comfort-related variables. Participants rate these items using a standardized scale, and the responses are analysed statistically to extract the EDFs most relevant to external ear comfort. Alongside the physical comfort dimensions, the surveys of this book also collect information on usage motivations and contextual factors. These elements are retained for subsequent analysis and are revisited in Chap. 7, where their contribution to dynamic comfort perception is examined.

2.3.1 Ethics Approval

The studies of this chapter performed in line with the principles of the Declaration of Helsinki. Informed consent was obtained from all participants. The studies of this chapter were approved by the Ethics Committee of Northwestern Polytechnic University (No. IRB No. 201502024, No. 202002024).

2.3.2 Comfort Issues and Descriptive Variables

A semi-structured interview approach is adopted in this book to explore comfort experiences with ear-related wearables. A total of 362 users of relevant products participate in discussions about issues influencing their perceived comfort. The interviews focus on product design types, fitting location or area types, daily usage duration, usage scenarios and contexts, usage motivations, and comfort-related experiences. From the analysis of these discussions, comfort issues, which are defined as fitting locations or areas associated with discomfort, together with the corresponding descriptive variables, are identified. To ensure that potential variables are not overlooked from the backend, or product-oriented, perspective, professionals also should contribute to the discussions (including five product designers and five product engineers, each with more than 5 years of industry experience). The semi-structured interviews extend over a seven-month period in this book, providing both depth and breadth of data for the extraction of ergonomic design factors.

Sample Size

A reasonable sample size is essential to effectively estimate the characteristics of a target population [40, 41]. The minimum number of participants required to estimate the mean characteristics from samples of convenience can be calculated as:

$$N = \frac{z_{\frac{\alpha}{2}}^{2} \beta(1-\beta)}{\delta^{2}} \tag{2.7}$$

where, N is the number of participants; $z_{\frac{\alpha}{2}}$ is the t-distribution statistic equivalent to the normal distribution at significance level $\alpha = 0.05$ (two-tailed interval), $\alpha = 0.05$ (95% confidence interval), $z_{0.025}=1.96$; β is the Type II error probability, $1 - \beta$ represents test power; δ is the expected estimation error. The minimum sample size for convenience samples is estimated to be 30 [41]. However, convenience samples require a high level of similarity among individual characteristics, making it difficult to control actual sampling error.

Changes in the external ear wearing comfort depend on variations in the EDFs, which are derived from diverse user anthropometric attributes. Furthermore, ear-related wearables involve multiple fitting locations/areas and several auricular substructures. The determination of the minimum sample size should conform to the requirements specified in ISO 15535:2012 [42] for the establishment of anthropometric databases. Specifically:

$$N = \left(\frac{z_{0.025} \times CV}{a}\right)^{2} \times 1.534^{2} \tag{2.8}$$

$$n \geq \frac{N-1}{2} \tag{2.9}$$

where, N presents the number of participants; $z_{0.025}$ presents the t-distribution statistic equivalent to the normal distribution at significance level $\alpha = 0.05$ (two-tailed interval), $z_{0.025}=1.96$; CV is the coefficient of variation; a is the required relative precision (percentage), recommended as 1%; SD is the standard deviation of anthropometric data; $\overline{x}$ is the mean of anthropometric data.

In practice, Eq. (2.8) can be further simplified as:

$$N \geq \left(3.006 \times \frac{CV}{a}\right)^2 \tag{2.10}$$

$$n \geq \frac{N-1}{2} \tag{2.11}$$

where n represents the minimum number of participants for a single sex in mixed-sex samples.

As no dedicated external ear anthropometric database or national standard for the Chinese population exists prior to the preparation of this book, the minimum sample size is determined through an indirect estimation method based on head-face dimensions proposed by Farkas [36] and Ball [40]. According to the data provided in GB/T 2428–1998 [43] (Head-Face Dimensions of Adults, Chinese Population), the coefficient of variation reaches its maximum for head circumference ($CV_{max} \approx 5.5\%$). Under a 95% confidence interval and 1% relative precision, the estimated minimum sample size required for characterizing the external ear anthropometry of the Chinese population is 274 ($N \geq 273.34$). For mixed-sex samples, at least 137 male and 137 female participants are needed ($n \geq 136.5$). According to Zhao et al. [37], the latest age at which the external ears reach maturity in Chinese population is 14 years. Accordingly, the minimum age for participants in this study is set at 15 years. For participants under the age of 18, consent from legal guardians is required, and participation must be accompanied by the guardians.

Based on Sect. 2.3.1 regarding auricular cartilage framework, individuals with congenital auricular malformations, external trauma, corrective surgery history, or maxillofacial abnormalities should be excluded. Participants with a history of auricular piercings (e.g., lobe studs/rings/chains or cartilage studs/rings/chains) should also be excluded, due to potential disruptions to auricular cartilage and skin integrity [12]. To avoid potential confounding effects of genetic predispositions or behavioural similarities, participants with family relations (direct or collateral kinship within three generations) to already recruited subjects should be excluded.

Recruitment Locations and Methods

To minimize the influence of geographical distribution patterns and genetic predispositions on auricular anthropometric features, participants should be recruited from multiple cities across a country, with their province/state of origin recorded. The criteria for sampling city selection are: priority to transportation hub cities with high population mobility, ensuring a diverse representation; alignment with the population density distribution of a country.

According to the National Census of China (2010 and 2020), approximately 95% of the Chinese population resides southeast of the Hu Huanyong Line, which covers 43% of China's land area [44]. Sampling preparation for this book began prior to the Seventh National Census (2020). Moreover, traditional sampling locations used for establishing national anthropometric standards in China should also be considered. Taking these considerations into account, the sampling framework includes eight cities that represent the geographical and demographic diversity of China: Shenyang in the northeast, Tianjin in the north, Xuzhou and Shanghai in the east, Fuzhou in the southeast, Guangzhou in the south, Xi'an in the central-west, and Lanzhou in the west (Fig. 2.10). Except for Lanzhou, the other seven cities are located southeast of the Hu Huanyong Line. Recruitment for the semi-structured interviews, as well as for subsequent studies presented in this book, is carried out through a combination of online and in-person methods. These methods include the use of mobile social applications, online communities, campus posters, and public on-site recruitment. A summary of the participant recruitment criteria used in this book is presented in Table 2.2.

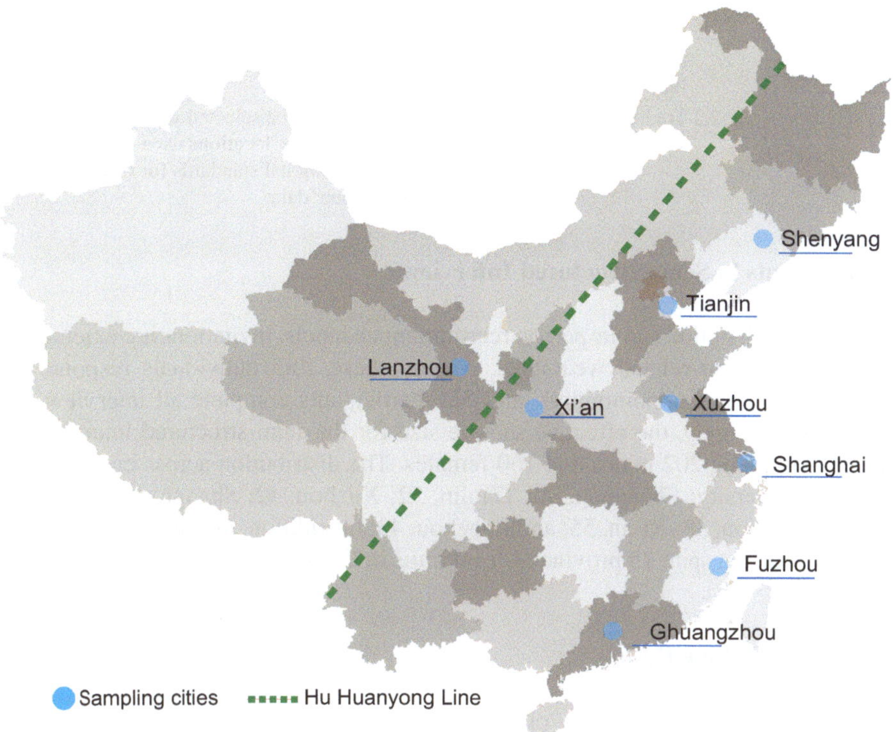

Fig. 2.10 Sampling cities

Table 2.2 Recruitment criteria for subjects participated used in this book

Item	Sub-item	Description or explanation
1. Recruitment criteria	1.1 Minimum sample size	The minimum sample size is 274, with at least 137 male and 137 female participants in the mixed-gender sample
	1.2 Age	The minimum age for recruited participants is 15 years
	1.3 External ear health status	Participants must have no history of congenital malformations, trauma/injury, corrective surgery, or maxillofacial abnormalities
	1.4 History of ear piercing	Participants must have no history of ear piercing
	1.5 Family relationships	Participants must not have family relationships with individuals already recruited
	1.6 Kinship relations	Participants must not be within three generations of direct or collateral kinship with individuals already recruited
2. Sampling sites	2.1 Transportation hub cities	Due to population mobility, participants' native provinces of origin should present a diverse distribution pattern
	2.2 Distribution along the Hu Huanyong line	The geographical distribution of participants should reflect the population density distribution characteristics of China
	2.3 Traditional sampling sites	Sampling sites should be selected with reference to the traditional sampling locations used in the establishment of national standards for Chinese population attribute data

Participants in Semi-Structured Interviews

Through both online and in-person recruitment channels, invitations are extended to 600 users of ear-related wearables. Among these, 485 individuals respond and arrange interview appointments, and 362 participants complete all interview sessions. Accordingly, the effective sample size for the semi-structured interviews is 362, comprising 202 males and 160 females. The distribution across cities reflects regional diversity: Shenyang, 40; Tianjin, 52; Xuzhou, 42; Shanghai, 46; Fuzhou, 46; Guangzhou, 38; Xi'an, 55; and Lanzhou, 43. In addition, the participants' native places of origin span 25 provincial-level units in China (Fig. 2.11).

Interview Outline Design

The semi-structured interview outline consists of (1) basic demographic information about the participants, (2) comfort-related questions on ear-related wearables, and (3) interview-related information (Table 2.3). The demographic information covers key variables such as age, gender, height, weight, province of origin, education level, and occupation. The interview-related information records the format of

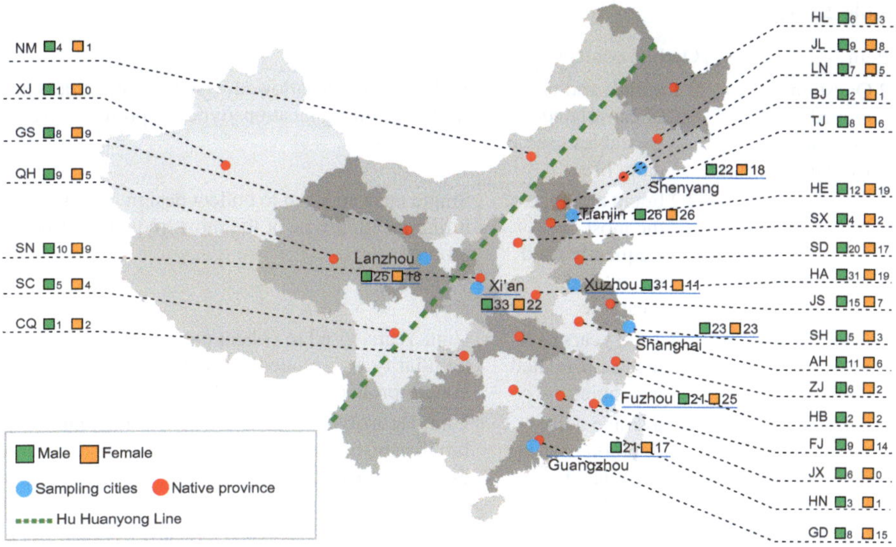

Fig. 2.11 Distribution of participants across sampling locations and native provinces in the semi-structured interview

Table 2.3 Outline and descriptions of the semi-structured interview for wearing comfort experience of human external ear

Item	Outline content	Description
1 Basic information	1.1 Age	Unit: Years
	1.2 Sex	Male/female
	1.3 Height	Unit: Cm
	1.4 Weight	Unit: Kg
	1.5 Native Province	Recorded according to the codes for the administrative divisions of the People's republic of China
	1.6 Education level	(1) high school and below; (2) associate degree; (3) bachelor's degree; (4) graduate (master/doctor)
	1.7 Occupation	Classified based on the online occupational classification document of the occupational skill appraisal Centre, Ministry of Civil Affairs of the People's republic of China, and GB/T 4754–2017 (industrial classification for National Economic Activities)

(continued)

Table 2.3 (continued)

Item	Outline content	Description
2 Questions related to the external ear wearing comfort	2.1 brands and types of ear-related wearables used in daily life	Multiple answers allowed. Used to collect design types and fitting locations/types of products used in daily life.
	2.2 brand and type of the most frequently used ear-related wearable	Single answer. Used to collect design type and fitting location/type of the most frequently used product
	2.3 Total years of using ear-related wearables	Unit: Years. Used to collect participants' familiarity with related products and their human-product interaction history
	2.4 average daily duration of using ear-related wearables	Unit: Hours/day. Used to collect participants' familiarity, human-product interaction history, and duration of usage motivation
	2.5 Main reasons for daily use of ear-related wearables	Work-related needs, assistive/hearing enhancement needs, entertainment-related needs, etc. multiple answers allowed. Used to collect types of usage motivations
	2.6 the most direct reason for daily use of ear-related wearables	Work-related needs, assistive/hearing enhancement needs, entertainment-related needs, etc. single answer. Used to collect types of usage motivations
	2.7 usage scenarios/ contexts of ear-related wearables	Used to collect scenarios/contexts associated with usage motivations
	2.8 Most frequent usage scenarios/ contexts of ear-related wearables	Used to collect the most frequent scenarios/ contexts associated with usage motivations
	2.9 descriptions of comfort experience issues with ear-related wearables	Fully open-ended discussion question. Used to collect specific descriptive variables of factors influencing external ear wearing comfort
3 Interview-related information	3.1 Interview format	(1) Online; (2) in-person
	3.2 Interview location	(1) Shenyang; (2) Tianjin; (3) Xuzhou; (4) Xi'an; (5) Shanghai; (6) Fuzhou; (7) Guangzhou; (8) Lanzhou; (9) zoom or other instant video communication/conferencing software
	3.3 Interview duration	Actual duration of interview and discussion, recorded in minutes. Planned duration per participant: 30 min

the interview (online or in-person), the location, and the duration. The comfort-related questions are designed to address specific aspects of participants' experiences and are organized as follows:

Q2.1 and Q2.2: product design type and fitting location/area types in daily use;

1. Q2.3 and Q2.4: user familiarity, human-product interaction history, and usage duration;
2. Q2.5 and Q2.6: user motivations for product use;
3. Q2.7 and Q2.8: usage scenarios/contexts;
4. Q2.9: variables related to comfort-affecting factors of ear-related wearables.

Data Collection and Analysis

To minimize inter-observer variability, all interview sessions and data collection are carried out by two designated researchers. Researcher A takes responsibility for conducting the interviews, facilitating discussions, and preparing paper-based notes, while Researcher B manages non-paper records such as audio, video, and screen recordings. Paper records are made with pen and paper, whereas non-paper records rely on digital tools, including voice recorders, video cameras, or recording modules embedded in instant communication and conferencing software. Prior to participation, all interviewees sign an informed consent form. For participants under 18 years of age, consent is obtained from legal guardians, who also accompany them during the sessions. When permitted by participants or guardians, both audio and video recordings are collected; otherwise, only audio data are obtained. Due to scheduling conflicts, 36 of the 362 participants could not attend in-person sessions and therefore completed their interviews online. Each session is planned to last approximately 30 min; in practice, actual durations ranged from 24 to 59 min, with a mean of 43.40 min (SD = 8.05). Data analysis is conducted using MATLAB R2018a, applying descriptive statistical procedures.

The participants range in age from 17 to 75 years, and their demographic distribution is summarized in Table A.1. In terms of occupational categories, the proportions of students, retired individuals, and others are 19.89%, 14.64%, and 65.47%, respectively. More detailed demographic data across different age groups is provided in Tables A.2 and A.3.

According to Tables 2.4 and 2.5, discussions with participants on Q2.1, Q2.2, Q2.5, and Q2.6 indicate that ear-muff and ear-bud type products are preferred by a higher proportion of users. More than 40% of participants describe themselves as being in a "passive" state of long-term use imposed by work requirements, particularly in industries such as manufacturing, construction, transportation, postal, and telecommunication services.

Users selecting ear-cup type products primarily seek effective noise reduction during work or exercise. By contrast, ear-hook or band-type products are reported as the least frequently used, largely due to prominent comfort issues in their current designs. The number of participants driven by entertainment-related motivations is

Table 2.4 Distribution of participants across product types in the semi-structured interview

	Ear-cup		Ear-muff		Ear-hook/ band		Ear-bud/plug	
Product type	N	%	N	%	N	%	N	%
Product types used in daily life	162	44.75	282	77.90	62	17.12	263	72.65
Most frequently used product type	85	23.48	135	37.29	17	4.70	125	35.16

Table 2.5 Distribution of participants across usage purposes in the semi-structured interview

	Work-related purposes		Assistive/hearing purposes		Entertainment-related purposes	
Usage purpose	N	%	N	%	N	%
Primary reasons for daily product use	275	75.97	23	6.35	305	84.25
Most direct reason for daily product use	155	42.82	23	6.35	184	50.83

Table 2.6 Distribution of participants across total years and average daily usage duration in the semi-structured interview

Total years of daily product use	N	%	Average daily usage duration	N	%
<5 years	2	0.55	<1 h/day	5	1.38
5 years	26	7.18	1 h/day	17	4.70
6 years	48	13.26	2 h/day	23	6.35
7 years	64	17.68	3 h/day	47	12.98
8 years	76	20.99	4 h/day	184	50.83
9 years	54	14.92	4–8 h/day	63	17.40
10 years	40	11.05	>8 h/day	23	6.35
>10 years	52	14.36			

roughly comparable to those driven by work-related needs. Entertainment-related use is especially common among students, retirees, and professionals in fields such as education, culture and the arts, broadcasting, new media and online communication, scientific research and technical services, finance, insurance, government, and social organizations. It is noteworthy that all participants report having used at least two different design types of related products in their daily lives. According to Table 2.6, discussions of Q2.3 and Q2.4 further show that only two participants (both students) had less than 5 years of cumulative product usage, while more than 70% of participants report an average daily usage duration exceeding 4 h.

Fitting Locations/Areas with Comfort Issues

In discussions related to Q2.9, participants consistently point out that ear-related wearables present notable comfort challenges. Among those whose primary motivation for daily use is work-related (155 individuals) or lifestyle-related (184 individuals), all indicate that their current products suffer from comfort-related design problems and emphasize the need for optimization. Participants whose main motivation is assistive or hearing enhancement purposes (23 individuals) also acknowledge the presence of comfort issues; however, they unanimously stress that functional performance takes precedence over comfort. From these discussions, seven fitting locations or areas emerge as critical sources of discomfort (Table 2.7): (1) the outer contour of the auricle, (2) the anterior surface of the auricle, (3) the upper and posterior regions of the otobasion, (4) the concha, (5) the tragus, (6) the entrance of the external acoustic meatus, and (7) the canal sections and bends of the external acoustic meatus. Across these sites, the primary physical comfort problems arise from mismatches between product dimensions and the corresponding external ear dimensions, as well as from localized compression of the skin and the underlying anatomical structures at the product-ear interface.

Descriptive Variables for Measurement

According to participants' discussions on Question 2.9, a preliminary set of descriptive variables for measurement is identified. These variables include: (1) the size of the auricular outer contour, (2) the size of the otobasion, (3) the size of the concha, (4) the size of the tragus, (5) the size of the cavum concha, (6) the size of the auricular superior protrusion, (7) the size of the auricular posterior protrusion, (8) the size of the entrance of the external acoustic meatus, (9) the size of the bends of the external acoustic meatus, and (10) the size of the sections of the external acoustic meatus. Following further consultation with product designers and engineers, the earlobe dimension is added as a compensatory variable. This addition addresses potential comfort concerns associated with the fitting of ear-cup products. For instance, the contact between the product and the auricle may result in folding of the earlobe during use. The final set of descriptive variables, together with corresponding participant statements, is summarized in Table 2.8.

2.3.3 EDFs of Wearing Comfort

A structured questionnaire survey is employed in this book to examine the ergonomic design factors (EDFs) that are associated with external ear comfort. A total of 859 users of ear-related wearables is invited to participate. The questionnaire items draw directly on the fitting locations and descriptive variables identified in Sect. 2.3, ensuring continuity between qualitative exploration and quantitative validation. By analysing the responses, the EDFs most relevant to wearing comfort of the external ear are extracted. The survey is conducted over a four-month period.

Table 2.7 Fitting locations/areas with comfort issues by most frequently used product type in the semi-structured interview

Most frequently used product type	N of participants	Fitting locations/ areas with comfort issues	Specific location	Transcribed comfort issue description
Ear-cup	85	Auricle	Outer contour	Product dimensions did not fit to the outer contour of the auricle or the upper and posterior sides of the otobasion; compression of the anterior auricular skin by the product-ear contact surface caused discomfort
			Anterior surface of the auricle	
			Upper and posterior sides of the otobasion	
Ear-muff	135	Auricle	Anterior surface of the auricle	Compression of the anterior auricular skin by the product-ear contact surface caused discomfort
Ear-hook/ band	17	Auricle	Upper and posterior sides of the otobasion	Product dimensions did not fit to the upper and posterior sides of the otobasion
Ear-bud/ plug	125	Auricle	Concha	Product dimensions did not fit to the concha; compression of the anterior Conchal skin by the product-ear contact surface caused discomfort
			Tragus	Product dimensions did not fit to the tragus; compression of the tragus by the product-ear contact surface caused outward expansion and discomfort; compression of the inner tragus skin by the product-ear contact surface caused discomfort
		External acoustic meatus (EAM)	EAM entrance	Product dimensions did not fit to the entrance of the EAM; compression of the EAM entrance skin by the product-ear contact surface caused discomfort
			EAM sections and bends	Product dimensions and product morphology did not fit to the canal sections and bends of the EAM

Participant Recruitment

Both online and in-person recruitment are employed across eight sampling cities. Invitations are distributed to 1261 users of ear-related wearables, of whom 963 agree to participate and schedule sessions. Ultimately, 859 participants complete the questionnaire, providing a valid sample size. Among them, 418 are male and 441 are female, meeting the requirements outlined in Table 2.2. As indicated in Tables 2.4 and 2.6, participants are required to have experience with all major design types of related products within the past 5 years and to possess long-term usage experience with at least two types. The final distribution of participants

Table 2.8 Descriptive variables for measurement

No.	Descriptive variable	Source of variable	Transcribed description
1	Auricular contour size	Interview results	Fitting of ear-cup products caused folding of the auricular contour; products that were too large or too small could not be used
2	Otobasion size	Interview results	Fitting of ear-cup and ear-hook/band products caused folding of the otobasion; products that were too large or too small could not be used
3	Concha size	Interview results	Fitting of ear-bud/plug and ear-hook/band products caused outward expansion of the concha; products that were too large or too small could not be used
4	Tragus size	Interview results	Fitting of ear-bud/plug and ear-hook/band products caused outward expansion of the tragus; products that were too large or too small could not be used
5	Cavum concha size	Interview results	Fitting of ear-bud/plug and ear-hook/band products caused outward expansion of the cavum concha; products that were too large or too small could not be used
6	Earlobe size	Compensatory variable	Fitting of ear-cup products caused folding of the earlobe; products that were too large or too small could not be used
7	Auricular superior protrusion size	Interview results	Fitting of ear-cup and ear-hook/band products caused outward expansion of the auricular superior protrusion; products that were too large or too small could not be used
8	Auricular posterior protrusion size	Interview results	Fitting of ear-cup and ear-hook/band products caused outward expansion of the auricular posterior protrusion; products that were too large or too small could not be used
9	Entrance size of the external acoustic meatus (EAM)	Interview results	Fitting of ear-bud/plug products caused outward expansion of the EAM entrance cross-section; products that were too large or too small could not be used
10	Bend size of the external acoustic meatus (EAM)	Interview results	Fitting of ear-bud/plug products caused enlargement of the EAM bend angle; products that were too large or too small could not be used
11	Section size of the external acoustic meatus (EAM)	Interview results	Fitting of ear-bud/plug products caused outward expansion of the EAM section cross-section; products that were too large or too small could not be used

across the eight locations is as follows: Shenyang, 82; Tianjin, 105; Xuzhou, 107; Shanghai, 107; Fuzhou, 113; Guangzhou, 105; Xi'an, 121; and Lanzhou, 119. Their native provinces extend across 30 provincial-level administrative regions in China. The geographical distribution of participants is illustrated in Fig. 2.12.

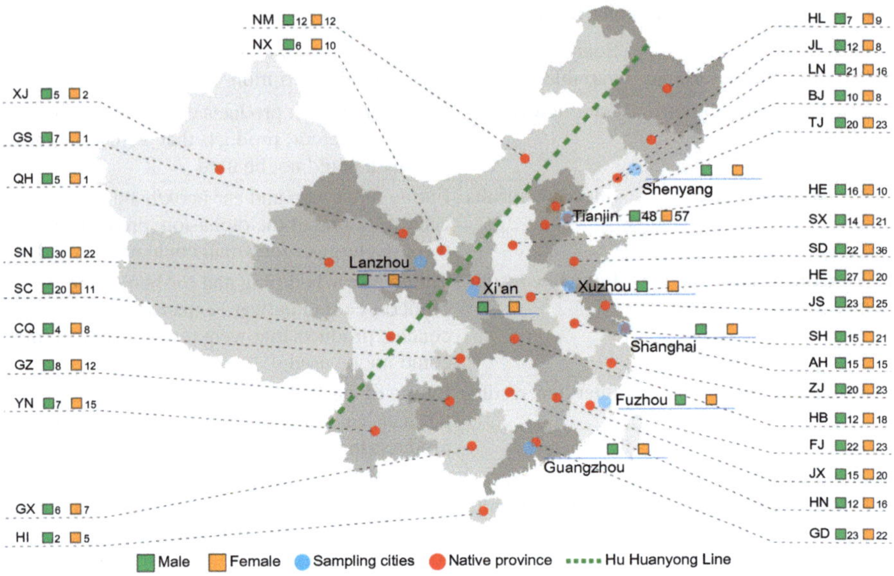

Fig. 2.12 Distribution of participants across sampling locations and native provinces in structured questionnaire survey

Questionnaire Design

The structured questionnaire consists of three main sections: (1) basic demographic information, including age, sex, height, weight, and native province; (2) a set of 11 questions addressing experiences of wearing comfort; and (3) survey-related information such as whether the data are collected online or in person at a sampling location. The 11 comfort-related questions are designed to evaluate whether incorporating the descriptive variables into product design contributes to an improved comfort experience (Table 2.9). Responses are recorded using a five-point Likert scale, where a score of 1 represents a slight effect and a score of 5 represents a very strong effect.

Data Collection and Analysis

To minimize inter-observer variability, three researchers are assigned specific responsibilities throughout the process. Researcher A oversees questionnaire distribution, collection, and participant guidance including Q&A; Researcher B is responsible for data entry; and Researcher C conducts verification and completeness checks of the entered data.

The ergonomic design factors (EDFs) related to external ear comfort are extracted through factor analysis of the questionnaire data. The procedure follows standard practice:

Table 2.9 Structured questionnaire for wearing comfort experience of the human external ear (for non-customized products)

Participant Basic Information (ID:)		
Age:	Sex:	Height:
Weight:	Native province:	
Questions related to external ear wearing comfort		
Survey of descriptive variables for the external ear wearing comfort		
Please rate the following 11 items based on your personal experience of using related products. Record your score in the parentheses after each item. The scoring options are 1, 2, 3, 4, and 5 (1 = slight effect on improving external ear wearing comfort; 5 = very strong effect on improving external ear wearing comfort).		
1. Consideration of the auricular contour size in product design ()		
2. Consideration of the otobasion size in product design ()		
3. Consideration of the concha size in product design ()		
4. Consideration of the tragus size in product design ()		
5. Consideration of the cavum concha size in product design ()		
6. Consideration of the earlobe size in product design ()		
7. Consideration of the auricular superior protrusion size in product design ()		
8. Consideration of the auricular posterior protrusion size in product desig ()		
9. Consideration of the entrance size of the external acoustic meatus (EAM) in product design ()		
10. Consideration of the bend size of the external acoustic meatus (EAM) in product design ()		
11. Consideration of the section size of the external acoustic meatus (EAM) in product design ()		
Survey-related information		
Survey format:	Survey location:	Survey date:

1. First, perform Kaiser-Meyer-Olkin (KMO) test and Bartlett's test. Data were deemed suitable when KMO > 0.6 and Bartlett's $p < 0.05$ [45].
2. Then, factor analysis was conducted using varimax orthogonal rotation to maximize within-factor correlations.
3. Variables with factor loadings >0.4 were retained [46].

All statistical analyses are conducted in MATLAB R2018a under a 95% confidence level ($\alpha = 0.05$), and results with $p < 0.05$ are regarded as statistically significant.

Factors

The 859 participants range in age from 18 to 79 years, and their demographic characteristics are presented in Tables A.4 and A.5. As shown in Table 2.10, ten descriptive variables achieve factor loadings greater than 0.4, while otobasion dimensions are excluded from the analysis. At the current stage of design research, the contour of the auricular otobasion cannot be standardized; as a result, product designs generally do not attempt to conform to or accommodate this feature. Consequently,

Table 2.10 Physical ergonomic design factors of wearing comfort for the human external ear

	Factor and corresponding questionnaire item[c]		
	Factor 1 (F_{Aur})	Factor 2 (F_{EAM})	Factor 3 (F_{A-E})
Variable (factor loadings >0.4)[a,b]	Auricle shape-structure	External acoustic meatus (EAM) shape-structure	Auricle-EAM transitional structure
Auricular contour size	1		
Cavum concha size	5		
Tragus size	4		4
Concha size	3		
Earlobe size	6		
Auricular superior protrusion size	7		
Auricular posterior protrusion size	8		
Entrance size of EAM		9	9
Section size of EAM		11	
Bend size of EAM		10	
% variance contribution	37.95	28.35	10.55
Cumulative variance contribution %	37.95	66.3	76.85
Number of items	7	3	2

[a] Only variables with factor loadings greater than 0.4 are retained
[b] Rotation converged in 7 iterations
[c] Questionnaire item no. corresponds to Table 2.9, eigenvalue >1

participants express a relatively negative attitude toward the ergonomic significance of the otobasion.

The factor analysis results (KMO = 0.78, Bartlett's $p < 0.001$, cumulative variance contribution = 76.85%) grouped the 10 retained descriptive variables into three factors: (1) the auricle shape-structure factor (Factor 1, F_{Aur}); (2) the external acoustic meatus shape-structure factor (Factor 2, F_{EAM}); (3) the auricle-EAM transitional structure factor (Factor 3, F_{A-E}). The tragus size variable and EAM entrance size variable fall within the research domain of the transition structure between the auricle and the external acoustic meatus (EAM), which are key design elements of band-type and plug-type products. Hence, Factor 3 describes two descriptive variables associated with the transitional structure between the auricle and the external acoustic meatus (EAM). Among the ten descriptive variables, both the tragus size and the EAM entrance size are shared across factors, producing a hybrid classification in which overlapping variables bridge multiple structural domains. This overlapping character illustrates the anatomical and functional interdependence of auricular and canal structures in shaping comfort outcomes.

Finally, based on the cumulative variance contribution (Table 2.10), the static multi-factor model of wearing comfort perception (Eq. (2.1)) can be further expressed using the Factor Weighting Formula (Eq. (2.12)), yielding Eq. (2.13):

$$\omega_i = \frac{(\%of\ Variance)_i}{\sum_{i=1}^{n}(\%of\ Variance)_i} \tag{2.12}$$

$$P_{Fitting} = 0.49P_{Aur} + 0.37P_{EAM} + 0.14P_{A-E} \tag{2.13}$$

where, i presents an EDF; *% of Variance* is the variance contribution rate; ω presents the factor weight; m: presents the number of factors; P_{Aur} is the comfort perception related to F_{Aur}; P_{EAM} is the comfort perception related to F_{EAM}; P_{A-E} is the comfort perception related to F_{A-E}.

The shape-structure characteristics of the external ear provide a foundation for related design research and applications. Studies on the auricle shape and structure are relatively abundant; however, quantitative analyses of variables have typically focused on specific product types, resulting in a lack of systematic characterization research on the auricle itself. In contrast, studies on the shape and structure of the external acoustic meatus (EAM) remain limited. Due to the absence of clearly defined measurement variables, significant gaps still exist in the structural understanding of the EAM. Accordingly, the following chapters will first address the shape-structure characterization and comfort perception quantification of the auricle and the EAM. The specific objective is to establish a set of measurement variables representing the auricle and EAM shape and structure. It will enable systematic analysis and provide both descriptive outcomes and a quantitative basis for subsequent research on wearing comfort perception. On this basis, suitable measurement variables and deformation types should be identified for constructing deformation-comfort relationships. Finally, through actual measurement results and model fitting outcomes, this book aims to provide comfort thresholds and comfort perception limits for subsequent studies.

The auricle-EAM transitional structure (F_{A-E}) represents the primary fitting region for intrusive products such as plug-type and band-type devices. In the absence of defined descriptive variables and measurement methods, general understanding of the expansion of the auricle-EAM transitional structure remains limited. Building on the auricle shape-structure factor (F_{Aur}) and the EAM shape-structure factor (F_{EAM}), the following chapters will explore the relationship between transitional structure expansion and comfort perception effects of auricle-EAM transitional structure expansion. The objective is to characterize the auricle-EAM transitional structure and to quantify the perception of comfort associated with its expansion, thereby establishing comfort perception-effect relationships between F_{A-E} and changes in wearing comfort perception. Specifically, this book aims to: (1) define measurement variables for the F_{A-E} and its expansion, (2) design standardized instruments and measurement tools for assessing F_{A-E} expansion-related comfort perception, and (3) establish quantitative relationships between expansion variables and changes in comfort perception. These efforts could further allow the calculation of comfort thresholds and comfort perception limits, as well as assessment of the offset effects of functional comfort perception on physical comfort perception.

According to the theoretical models and the previously established interaction path model, the design mechanisms of external ear wearing comfort should also account for comfort loss associated with duration variations in specific usage scenarios or contexts. Hence, the subsequent chapters should employ self-reported data collected during the extraction of ergonomic design factors to construct quantitative relationships between usage motivations and comfort loss in external ear wearing comfort perception. Integrating the results and findings from this book, a static mechanism model (SMM) of external ear wearing comfort perception could be established. By synthesizing this static model with the comfort loss model, a dynamic mechanism model (DMM) could be developed for both design research and practical applications. The subsequent chapters will present the overall structure, application rules, and usage guidelines of these models. Through case studies, this book will further demonstrate practical methods for applying the comfort perception-effect model of external ear wearing comfort. It will also validate the model's usability and effectiveness by comparing computational results with participant comfort feedback.

2.4 Toward a Framework of Physical Comfort in Ear-Related Wearables

2.4.1 Establishing Theoretical Models of Comfort Perception

By developing both static and dynamic frameworks, the models reveal how mismatches between product parameters and ergonomic design factors (EDFs) shape comfort outcomes. Static multi-factor and single-factor models highlight dimensional variations, while dynamic models incorporate motivational variables such as activity type, duration, and usage scenario. These formulations reposition comfort as a definable interaction between human attributes and product parameters, rather than a purely subjective state.

The auricle and external acoustic meatus (EAM) were examined in terms of cartilage framework, skin innervation, and muscular distribution, each of which imposes opportunities and constraints for design. Auricular elasticity supports fixation but also sets safe deformation limits, while dense cutaneous innervation makes surface pressure a critical comfort determinant. Growth trajectories further highlight user diversity, as maturation ages differ across auricular dimensions and vary between sexes. Quantitative references for deformation thresholds provide designers with practical parameters for anticipating user discomfort.

2.4.2 Extraction of Ergonomic Design Factors

Building on theoretical and structural foundations, empirical methods for extracting EDFs are also presented. Through semi-structured interviews with 362 participants and structured questionnaires with 859 participants across eight Chinese cities, a robust and representative dataset was created. These investigations identified fitting locations prone to comfort issues, including auricular contour, tragus, and EAM bends, and translated them into descriptive variables. Statistical methods such as KMO tests, Bartlett's test, and factor analysis grouped variables into three coherent categories: auricular shape-structure, EAM shape-structure, and auricle-EAM transitional structure. This taxonomy bridges user experiences with measurable anthropometric dimensions, offering a dual perspective on comfort.

The methodological rigor of the research further strengthens its contributions. By combining qualitative interviews with quantitative surveys, the study ensured that EDFs were simultaneously grounded in user experiences and validated through statistical generalization. Sample size requirements adhered to ISO 15535 standards, while explicit exclusion criteria (e.g., congenital malformations, piercings, kinship relations) safeguarded data quality. These measures underscore the importance of combining anthropometric precision with ergonomic relevance when building comfort models.

2.4.3 Key Insights and Practical Implications

Several critical insights emerge from the findings. Comfort is a cumulative outcome of multiple EDFs acting together rather than a single parameter such as size or pressure. Transitional structures between the auricle and EAM are particularly sensitive zones, posing ergonomic challenges for intrusive devices like earplugs and in-ear earphones. Moreover, variability in comfort perception across individuals highlights the need for inclusive design strategies that consider sex-related differences, age variation, and bilateral asymmetry. The integration of dynamic comfort modelling is especially significant, as it acknowledges that comfort must be understood within real-world scenarios involving motion, extended wear, and diverse environments.

For product designers, the EDF framework offers a roadmap for transforming anthropometric data into actionable design criteria. For ergonomists, it provides predictive tools to establish thresholds and exposure limits. For researchers, the empirical findings create a foundation for further work on deformation-comfort relationships and threshold quantification. The potential for adaptive or personalized wearables that respond to changes in activity and motivation reflects a forward-looking application of these insights.

Competing Interests This study is by "the Fundamental Research Funds for the Central Universities" (NPU Grant No. D5000250035, SEU Grant No. 2242024S30013). Additional support comes from the Northwestern Polytechnical University. The authors have no conflicts of interest to declare that are relevant to the content of this chapter.

References

1. Doutres, O., Sgard, F., Terroir, J., Perrin, N., Jolly, C., Gauvin, C., & Negrini, A. (2019). A critical review of the literature on comfort of hearing protection devices: Definition of comfort and identification of its main attributes for earplug types. *International Journal of Audiology, 58*(12), 824–833. https://doi.org/10.1080/14992027.2019.1646930
2. Doutres, O., Sgard, F., Terroir, J., Perrin, N., Jolly, C., Gauvin, C., & Negrini, A. (2020). A critical review of the literature on comfort of hearing protection devices: Analysis of the comfort measurement variability. *International Journal of Occupational Safety and Ergonomics, 28*(1), 447–458. https://doi.org/10.1080/10803548.2020.1772546
3. Mansfield, N., Naddeo, A., Frohriep, S., & Vink, P. (2020). Integrating and applying models of comfort. *Applied Ergonomics, 82*, 102917. https://doi.org/10.1016/j.apergo.2019.102917
4. Wang, M., Fan, H., Yu, S., Zhao, X., Wang, L., Li, W., Wang, L., Yu, M., Chu, J., Zhang, S., & Chen, D. (2022). Effects of variations in the tragus expansion angle on physical comfort for in-ear wearables. *Ergonomics, 65*(10), 1352–1372. https://doi.org/10.1080/00140139.2022.2032377
5. Waters, T. R., Putz-Anderson, V., Garg, A., & Fine, L. J. (1993). Revised NIOSH equation for the design and evaluation of manual lifting tasks. *Ergonomics, 36*(7), 749–776. https://doi.org/10.1080/00140139308967940
6. Demir, E., Desmet, P. M. A., & Hekkert, P. (2009). Appraisal patterns of emotions in human-product interaction. *International Journal of Design, 3*(2), 41–51. http://www.ijdesign.org/index.php/IJDesign/article/view/522
7. Ahmed, S., Zhang, J., & Demirel, O. (2018). Assessment of types of prototyping in human-centred product design. In *Lecture notes in computer science* (Vol. 10917, pp. 3–18). Springer. https://doi.org/10.1007/978-3-319-91397-1_1
8. Naddeo, A., Cappetti, N., Vallone, M., & Frohriep, S. (2014). New trend line of research about comfort evaluation: Proposal of a framework for weighing and evaluating contributions coming from cognitive, postural and physiologic comfort perceptions. In *Advances in social and organizational factors* (pp. 503–515). CRC Press.
9. Schifferstein, H. N. J., & Hekkert, P. (2011). *Product experience*. Elsevier.
10. Standring, S. (2016). *Gray's anatomy: The anatomical basis of clinical practice* (41st ed.). Elsevier.
11. Shah, P., Luximon, Y., & Luximon, A. (2022). Measurement of soft tissue deformation at discomfort and pain threshold in different regions of the head. *Ergonomics, 65*(9), 1261–1276. https://doi.org/10.1080/00140139.2022.2028016
12. Ahmed, A. A., & Omer, N. (2015). Estimation of sex from the anthropometric ear measurements of a Sudanese population. *Legal Medicine, 17*(5), 313–319. https://doi.org/10.1016/j.legalmed.2015.03.002
13. Roebuck, J. A. (2015). Fixing flaws in engineering anthropometry, ears and all. In *Proceedings of the human factors and ergonomics society annual meeting* (Vol. 59, pp. 1414–1418). SAGE. https://doi.org/10.1177/1541931215591307
14. Fan, H., Yu, S., Chu, J., et al. (2019). Anthropometric characteristics and product categorization of Chinese auricles for ergonomic design. *International Journal of Industrial Ergonomics, 69*, 118–141. https://doi.org/10.1016/j.ergon.2018.11.002

15. Kurosawa, T., Taniguchi, T., Momose, T., Takeda, T., & Yamashita, Y. (2019). Simultaneous measurement of ear canal movement, electromyography of the masseter muscle and occlusal force for earphone-type occlusal force estimation device development. *Sensors, 19*(15), 3357. https://doi.org/10.3390/s19153441
16. Lee, W., Jung, H., Bok, I., & Kim, J. (2016). Measurement and application of 3D ear images for earphone design. In *Proceedings of the human factors and ergonomics society annual meeting* (Vol. 60, pp. 1053–1057). SAGE. https://doi.org/10.1177/1541931213601244
17. Harris-Adamson, C., Eisen, E. A., Kapellusch, J., Garg, A., Hegmann, K. T., Thiese, M. S., & Rempel, D. M. (2015). Biomechanical risk factors for carpal tunnel syndrome: A pooled study of 2474 workers. *Occupational and Environmental Medicine, 72*(1), 33–41. https://doi.org/10.1136/oemed-2014-102378
18. Yan, Y., Fan, H., Li, Y., et al. (2021). Applying wearable technology and a deep learning model to predict occupational physical activities. *Applied Sciences, 11*(20), 9436. https://doi.org/10.3390/app11209436
19. Yan, Y., Fan, H., Li, Y., et al. (2020). Prediction of occupational physical activities using inertial measurement units and deep learning models. In *Proceedings of the human factors and ergonomics society annual meeting* (Vol. 64, pp. 879–883). SAGE. https://doi.org/10.1177/1071181320641206
20. Zhang, S., He, W., Chen, D., et al. (2019). A dynamic human reliability assessment approach for manned submersibles using PMV-CREAM. *International Journal of Naval Architecture and Ocean Engineering, 11*(2), 782–795. https://doi.org/10.1016/j.ijnaoe.2019.03.002
21. Zhang, S., He, W., Chen, D., et al. (2019). Thermal comfort analysis based on PMV/PPD in cabins of manned submersibles. *Building and Environment, 148*, 668–676. https://doi.org/10.1016/j.buildenv.2018.10.033
22. Harris-Adamson, C., Eisen, E. A., Dale, A. M., Evanoff, B., Hegmann, K. T., Thiese, M. S., & Rempel, D. M. (2013). Personal and workplace psychosocial risk factors for carpal tunnel syndrome: A pooled cohort study. *Occupational and Environmental Medicine, 70*(8), 529–537. https://doi.org/10.1136/oemed-2013-101365
23. Kaljun, J., & Dolšak, B. (2012). Ergonomic design knowledge built in the intelligent decision support system. *International Journal of Industrial Ergonomics, 42*(1), 162–171. https://doi.org/10.1016/j.ergon.2011.11.009
24. Dueñas, L., Balasch Bernat, M., Mena del Horno, S., & Martínez-Aldao, D. (2016). Development of predictive models for the estimation of the probability of suffering fear of falling and other fall risk factors based on posturography parameters in community-dwelling older adults. *International Journal of Industrial Ergonomics, 54*, 131–138. https://doi.org/10.1016/j.ergon.2016.05.009
25. Verhaert, V., Haex, B., de Wilde, T., Berckmans, D., Vandekerckhove, M., & Verbraecken, J. (2011). Ergonomics in bed design: The effect of spinal alignment on sleep parameters. *Ergonomics, 54*(2), 169–178. https://doi.org/10.1080/00140139.2010.538725
26. Yu, J. F., Lee, K. C., Wang, R. H., & Lee, C. C. (2015). Anthropometry of external auditory canal by non-contactable measurement. *Applied Ergonomics, 50*, 50–55. https://doi.org/10.1016/j.apergo.2015.01.008
27. Alexander, M., & Laubach, L. L. (1968). *Anthropometry of the human ear: A photogrammetric study of USAF flight personnel (Report No. AMRL-TR-68-23)*. Aerospace Medical Research Laboratories.
28. Barnes, J., Sabo, R. T., & Coelho, D. H. (2018). A novel method to measure the external auditory canal: Normative data and practical implications. *American Journal of Otolaryngology - Head and Neck Medicine and Surgery, 39*(2), 146–149. https://doi.org/10.1016/j.amjoto.2017.12.006
29. Fan, H., Yu, S., Wang, M., et al. (2021). Analysis of the external acoustic meatus for ergonomic design: Part I - Measurement of the external acoustic meatus using casting, scanning and rapid estimation approaches. *Ergonomics, 64*(5), 640–656. https://doi.org/10.1080/00140139.2020.1858188

30. Fan, H., Yu, S., Wang, M., et al. (2021). Analysis of the external acoustic meatus for ergonomic design: Part II - Anthropometric variations of the external acoustic meatus by sex, age and side in Chinese population. *Ergonomics, 64*(5), 657–670. https://doi.org/10.1080/00140139.2020.1867769
31. Wang, B., Dong, Y., Zhao, Y., Li, D., & Li, Y. (2011). Computed tomography measurement of the auricle in Han population of north China. *Journal of Plastic, Reconstructive & Aesthetic Surgery, 64*(1), 34–40. https://doi.org/10.1016/j.bjps.2010.03.053
32. Purkait, R., & Singh, P. (2007). Anthropometry of the normal human auricle: A study of adult Indian men. *Aesthetic Plastic Surgery, 31*(4), 372–379. https://doi.org/10.1007/s00266-006-0231-4
33. Berger, E. H., & Mitchell, I. (1989). Measurement of the pressure exerted by earmuffs and its relationship to perceived comfort. *Applied Acoustics, 27*(2), 79–88. https://doi.org/10.1016/0003-682X(89)90001-7
34. Park, M. Y., & Casali, J. G. (1991). An empirical study of comfort afforded by various hearing protection devices: Laboratory versus field results. *Applied Acoustics, 34*(3), 151–179. https://doi.org/10.1016/0003-682X(91)90082-P
35. Binderup, A. T., Arendt-Nielsen, L., & Madeleine, P. (2008). Pressure pain threshold mapping - A new imaging modality of muscle sensitivity to pain. In *2008 Annual IEEE student paper conference (AISPC), 25(c), 1-4*. https://doi.org/10.1109/AISPC.2008.4460549
36. Farkas, L. G., & Posnick, J. C. (1992). Growth and development of regional units in the head and face based on anthropometric measurements. *Cleft Palate-Craniofacial Journal, 29*(4), 301–302. https://doi.org/10.1597/1545-1569_1992_029_0301_gadrua_2.3.co_2
37. Zhao, S., Li, D., Liu, Z., et al. (2018). Anthropometric growth study of the ear in a Chinese population. *Journal of Plastic, Reconstructive & Aesthetic Surgery, 71*(4), 518–523. https://doi.org/10.1016/j.bjps.2017.10.010
38. Fu, F., & Luximon, Y. (2019). Fit and comfort perception on hearing aids: A pilot study. In A. Ahram et al. (Eds.), *International conference on human systems engineering and design: Future trends and applications* (pp. 360–364). Springer. https://doi.org/10.1007/978-3-030-02053-8_55
39. Alexander, K. S., Stott, D. J., Sivakumar, B., et al. (2011). A morphometric study of the human ear. *Journal of Plastic, Reconstructive & Aesthetic Surgery, 64*(1), 41–47. https://doi.org/10.1016/j.bjps.2010.04.005
40. Ball, R. (2009). 3-D design tools from the Size China project. *Ergonomics in Design, 17*(3), 8–13. https://doi.org/10.1518/106480409X12487281219931
41. Kirkwood, B. R., & Sterne, J. A. C. (2010). *Essential medical statistics* (2nd ed.). John Wiley & Sons.
42. International Organization for Standardization. (2012). *ISO 15535:2012 General requirements for establishing anthropometric databases*. ISO.
43. Standardization Administration of China. (1998). *GB/T 2428-1998 Head-face dimensions of adults, Chinese population*. China Standards Press.
44. Chen, M., Gong, Y., Li, Y., et al. (2016). Population distribution and urbanization on both sides of the Hu Huanyong Line: Answering the Premier's question. *Journal of Geographical Sciences, 26*(11), 1593–1610. https://doi.org/10.1007/s11442-016-1346-4
45. Kuo, C. C., Wang, M. J., & Lu, J. M. (2020). Developing sizing systems using 3D scanning head anthropometric data. *Measurement, 152*, 1–11. https://doi.org/10.1080/15459620500324727
46. Ahmadpour, N., Robert, J. M., & Lindgaard, G. (2016). Aircraft passenger comfort experience: Underlying factors and differentiation from discomfort. *Applied Ergonomics, 52*, 301–308. https://doi.org/10.1016/j.apergo.2015.07.029

Chapter 3
Shape-Structure Characterization of the External Ear

Abstract The external ear, consisting of the auricle and the external acoustic meatus (EAM), plays a critical role in sound transmission and serves as an essential interface for numerous ear-related devices, including hearing aids, earphones, and protective equipment. Despite its ergonomic importance, detailed anthropometric characterization of the external ear has historically been limited compared with head-face measurements. This chapter establishes a systematic framework for external ear shape-structure characterization through measurable variables, anatomical landmarks, and three-dimensional data acquisition. A total of 40 variables is defined, including 18 auricular measurements and 22 EAM measurements, covering linear, angular, circumference, and area parameters. These variables capture key morphological features such as auricular contour, conchal geometry, protrusion, canal entrance, bends, depth, and spatial orientation. Data were collected from 1120 participants across eight Chinese cities using non-contact 3D scanning combined with impression molding for the EAM. Statistical analyses, including ANOVA, correlation tests, and factor analysis, reveal significant effects of age, sex, and bilateral differences on external ear morphology. The proposed characterization provides an ergonomic foundation for designing ear-worn products, improving comfort thresholds, and enabling future individualized fitting applications.

Keywords Ear anthropometry · Auricle morphology · External acoustic meatus (EAM) · Ergonomic design variables · 3D shape-structure characterization

The external ear, comprising the auricle and the external acoustic meatus (EAM), represents a unique interface between the human body and its surrounding environment. Functionally, it channels sound waves toward the middle and inner ear while also serving as a site of interaction with numerous ear-related devices, ranging from industrial hearing protectors and medical hearing aids to modern consumer electronics such as earphones and virtual/augmented reality headsets. Unlike many other anatomical structures, the external ear is characterized by its delicate balance between cartilaginous resilience and soft-tissue vulnerability, making it particularly sensitive to issues of fit, pressure distribution, and long-term comfort. As such, the systematic characterization of external ear morphology is not only of anatomical

H. Fan et al., *Engineering Anthropometry of the External Ear for Human-Product Interaction*, https://doi.org/10.1007/978-981-92-2160-8_3

interest but also of direct ergonomic importance. Human physical comfort is the first step toward optimizing the global comfort issues of products [1, 2]. Numerous related studies have repeatedly emphasized a fundamental point: without appropriate dimensional design values, the comfort experience of users and any subsequent discussion cannot be meaningfully addressed [3, 4].

Despite this significance, ear anthropometry has historically received less comprehensive treatment than head-face measurements. Classical anthropometric studies emphasized head circumference, facial dimensions, and craniofacial indices, often treating the auricle and EAM as peripheral or secondary features. International and national standards provide valuable baselines for general ergonomics but fall short in representing the detailed morphology of the external ear. The absence of such detail becomes increasingly problematic as ear-related products evolve from simple accessories to high-performance systems requiring precise anatomical integration. For example, the sealing effectiveness of an earplug depends on correspondence with EAM cross-sections; similarly, the acoustic coupling of earphones is determined by the geometry of the concha and canal entrance. Without robust anthropometric foundations, product design risks remaining empirical, with comfort and performance depending largely on trial and error rather than systematic evidence.

In recent years, the need for high-resolution ear anthropometry has grown across multiple domains. Occupational health demands more accurate data to establish exposure limits for noise attenuation devices, particularly in industries with high sound levels and long wearing times. Medical practice requires detailed external ear and EAM models to improve reconstructive surgery, hearing aid fittings, and diagnostic imaging alignment. The consumer electronics industry increasingly relies on ear-conforming devices whose success depends on precise morphology. Virtual and augmented reality systems, for instance, not only rely on earphones but also on the physical stability of head-mounted devices that interact with auricular landmarks. Furthermore, with the global rise of personalized products and additive manufacturing, there is a growing expectation that ergonomic design will transition from population averages to individualized fittings. Therefore, it is essential to characterize the shape-structure of the external ear through measurable variables.

3.1 Characterization Variables for Measurement

3.1.1 Auricle

For the shape-structure characterization of the auricle, a total of 18 measurement variables is defined. Their detailed definitions and measurement methods are introduced in Sect. 3.2. As summarized in Sect. 2.3 (Table 2.8), the descriptive variables that constitute the auricle shape-structure involve the auricular contour, the cavum concha, the tragus, the concha, the lobule, and auricular protrusions at both the superaurale and tragal levels. Drawing on the findings of Stavrakos and

Ahmed-Kristensen [5, 6], Lee et al. [7], Roebuck [8–10], and Fan et al. [11], specific auricular landmarks within anatomical and engineering anthropometric frameworks are consistently regarded as critical reference points for ergonomic design and evaluation. These include the helix (together with the crus of helix), the concha (including both the cavum concha and cymba concha), the tragus, and the lobule (Fig. 2.8). The helix, for example, defines the general auricular contour apart from the lobule and provides the spatial division of the concha [12]. On this basis, an auricle variable set comprising 18 measurement variables is established: auricular length (AL), auricular width at postaurale (AWP), auricular width at the incisura anterior auris posterior (AWI), lobular length (LL), lobular width (LW), ear morphologic length (EML), auricular inclination angle (AIA), conchal length (CL), conchal width (CW), cavum concha length (CCL), cavum concha width (CCW), tragal height (TH), tragal length (TL), conchal depth (CD), ear protrusion at the superaurale level (EPS), ear protrusion at the tragal level (EPT), concho-mastoid angle at postaurale (CAP), and concho-mastoid angle at otobasion posterius (CAO). Among these, AWI, CCL, CCW, CAP, and CAO are introduced as newly defined characterization variables (Fig. 3.1). Of the 18 variables, AIA, CAP, and CAO are angular variables, while the remaining 15 are linear (Table 3.1).

Within this framework, the length variables AL, EML, and LL, together with the width variables AWP, AWI, and LW, characterize the overall size and shape of the auricular contour [13–21]. The orientation variable AIA represents the inclination angle of the auricular contour relative to the coronal plane of the head [13, 19, 21]. In the design of ear-worn devices, these variables are typically applied to determine the form of earcups and earmuffs in muff-type and cup-type products [4, 14]. Similarly, the length variables CL [7, 13], CCL, and TL [19, 20]; the width variables CW [16, 21, 22] and CCW; and the depth/height variables CD [18, 19] and TH [7, 13] define the size and morphology of the concha (including the cavum concha) and the tragus. These variables are particularly relevant to the design of components that fit within the conchal region [9–11, 13, 18, 23], improving the comfort of plug-type and band-type products. As emphasized by Jung et al. [4], Lee et al. [7], and Fan et al. [11], the cavum concha is the primary placement area for concha-fitting components. Accordingly, CCL and CCW are selected as the specific measurement variables representing the cavum concha.

The height variables EPS and EPT [13, 19], together with the angular variables CAP and CAO, serve to characterize the degree of auricular protrusion. As noted by Alexander et al. [19] and Purkait et al. [13], both linear and angular measures are necessary for a comprehensive assessment of auricular protrusion. Accordingly, CAP and CAO are introduced as supplementary angular variables. These parameters play an important role in shaping the cross-sectional forms of band-type product components. In addition, the variables AIA, EPS, EPT, CAP, and CAO collectively describe the relative spatial relationship between the auricle and the head, offering a more complete understanding of how auricular geometry interacts with wearable product design.

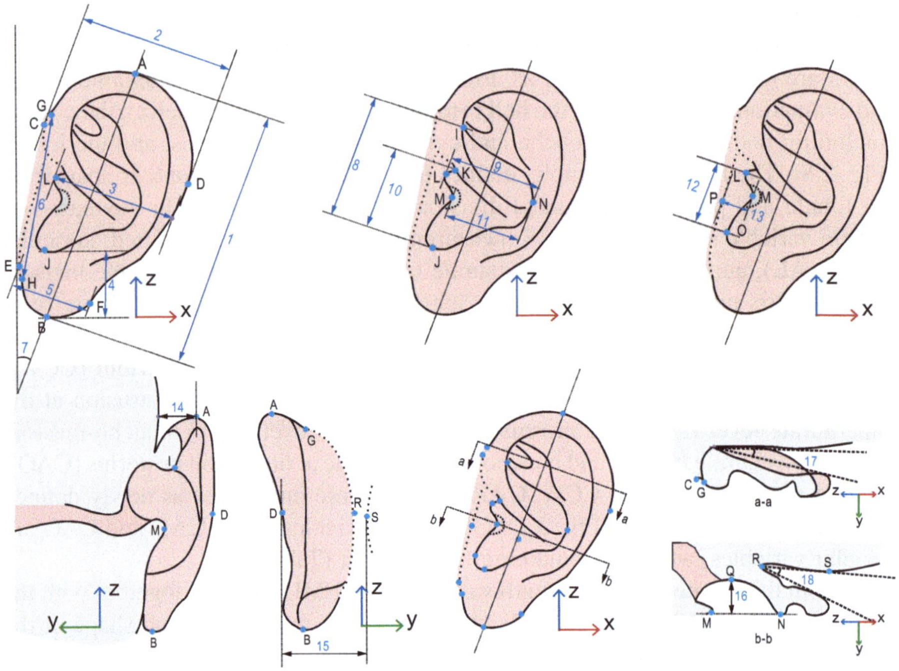

Fig. 3.1 Variables for auricular characterization:variables (1–18) include auricular length (1), auricular width at postaurale (2), auricular width at the incisura anterior auris posterior (3), lobular length (4), lobular width (5), ear morphologic length (6), auricular inclination angle (7), conchal length (8), conchal width (9), cavum concha length (10), cavum concha width (11), tragal length (12), tragal height (13), ear protrusion at superaurale-level (14), ear protrusion at tragal-level (15), conchal depth (16), concho-mastoid angle at postaurale (17), concho-mastoid angle at otobasion posterius (18); landmarks (A-S) include superaurale (A), subaurale (B), preaurale (C), postaurale (D), anterior lobule (E), posterior lobule (F), otobasion superius (G), otobasion inferius (H), anterior cymba concha (I), incisura intertragica inferior (J), superior cavum concha (K), incisura anterior auris posterior (L), tragion (M), posterior concha (N), lowest point on the lower border of tragus (O), cardinal point of tragus (P), medial concha (Q), otobasion posterius (R), anterior mastoid (S)

3.1.2 External Acoustic Meatus (EAM)

For the shape-structure characterization of the external acoustic meatus (EAM), a total of 22 measurement variables is defined. Their definitions and measurement methods are introduced in Sect. 3.2. As summarized in Sect. 2.3 (Table 2.8), the descriptive variables associated with the auricular structural factor involve the entrance of the EAM, its canal sections, and the canal bends. The anatomical features of the EAM are illustrated in Fig. 2.5. Based on the conclusions and practical findings of Yu et al. [24], as well as the structural characteristics described in Sect. 2.2, the region extending from the entrance to the second bend of the EAM provides the principal fitting support for plug-type and band-type products. Characterization

Table 3.1 Variables for auricular characterization

No.	Variable	Abbreviation	Variable type
1	Auricular length	AL	Linear
2	Auricular width at postaurale	AWP	Linear
3	Auricular width at the incisura anterior auris posterior	AWI	Linear
4	Lobular length	LL	Linear
5	Lobular width	LW	Linear
6	Ear morphologic length	EML	Linear
7	Auricular inclination angle	AIA	Angular
8	Conchal length	CL	Linear
9	Conchal width	CW	Linear
10	Cavum concha length	CCL	Linear
11	Cavum concha width	CCW	Linear
12	Tragal length	TL	Linear
13	Tragal height	TH	Linear
14	Ear protrusion at superaurale-level	EPS	Linear
15	Ear protrusion at tragal-level	EPT	Linear
16	Conchal depth	CD	Linear
17	Concho-mastoid angle at postaurale	CAP	Angular
18	Concho-mastoid angle at otobasion posterius	CAO	Angular

of this region establishes the ergonomic foundation for the design and evaluation of relevant products, particularly general-purpose and deep-protection hearing devices [24–27].

Accordingly, 22 measurement variables are selected to construct the EAM variable set, focusing on the structural region extending from the entrance to the second bend. The specific variables include: entrance length (EL), entrance width (EW), entrance circumference (EC), entrance area (EA), first bend length (FBL), first bend width (FBW), first bend circumference (FBC), first bend area (FBA), second bend length (SBL), second bend width (SBW), second bend circumference (SBC), second bend area (SBA), Section I depth (SD-I), Section II depth from first bend (SD-II-F), Section II depth from entrance (SD-II-E), entrance depth from geometric centre to medial concha (ED-C), entrance depth at posterior from medial concha (ED-P), azimuth angle of the geometric centre of the entrance (ACE), azimuth angle of Section I (AS-I), elevation angle of Section I (ES-I), azimuth angle of Section II (AS-II), and elevation angle of Section II (ES-II). Among these variables, EC, FBC, and SBC are classified as circumference variables; EA, FBA, and SBA as area variables; ACE, AS-I, ES-I, AS-II, and ES-II as angular variables; and the remainder as linear variables. The correspondence between descriptive and measurement variables is summarized in Table 3.2. Of the 22 variables, 18 are newly defined, with the exceptions of EL, EW, EC, and FBC (see Fig. 3.2).

Within this framework, EL, EW, EC, and EA are employed to describe the morphology and dimensions of the EAM entrance. FBL, FBW, FBC, and FBA define

Table 3.2 Variables of characterization for the external acoustic meatus

No.	Variable	Abbreviation	Variable type
19	Entrance length	EL	Linear
20	Entrance width	EW	Linear
21	Entrance circumference	EC	Circumference
22	Entrance area	EA	Area
23	First bend length	FBL	Linear
24	First bend width	FBW	Linear
25	First bend circumference	FBC	Circumference
26	First bend area	FBA	Area
27	Second bend length	SBL	Linear
28	Second bend width	SBW	Linear
29	Second bend circumference	SBC	Circumference
30	Second bend area	SBA	Area
31	Section I depth	SD-I	Linear
32	Section II depth from first bend	SD-II-F	Linear
33	Section II depth from entrance	SD-II-E	Linear
34	Entrance depth from geometric Centre to medial concha	ED-C	Linear
35	Entrance depth at posterior from medial concha	ED-P	Linear
36	Azimuth angle of geometric Centre of entrance	ACE	Angular
37	Azimuth angle of Section I	AS-I	Angular
38	Elevation angle of Section I	ES-I	Angular
39	Azimuth angle of Section II	AS-II	Angular
40	Elevation angle of Section II	ES-II	Angular

the morphology and dimensions of the first bend, while SBL, SBW, SBC, and SBA represent those of the second bend. These cross-sectional variables are commonly applied in determining the design cross-sectional shapes of in-ear components for plug-type and band-type products [28]. The three depth variables, SD-I, SD-II-F, and SD-II-E, describe the structural depth from the entrance to the second bend. They locate depth using the geometric centres of the entrance, first bend, and second bend cross-sections. This information is typically applied to determining the insertion length of in-ear components, particularly in the case of deep-protection hearing devices, which generally extend as far as the second bend [24]. The medial concha is employed to define the relative position of the EAM entrance and the auricle. Entrance-related positioning variables include ED-C, ED-P, and ACE. The accurate localization of the EAM entrance is essential for designing the connection between in-ear and conchal components, as well as for specifying the placement of acoustic elements such as speakers or microphones. Finally, the azimuth-angle variables (AS-I, AS-II) and elevation-angle variables (ES-I, ES-II) describe the positional relationships among canal sections. In combination with the ear horizontal plane (see Sect. 2.2), these variables define the relative spatial orientation of the EAM

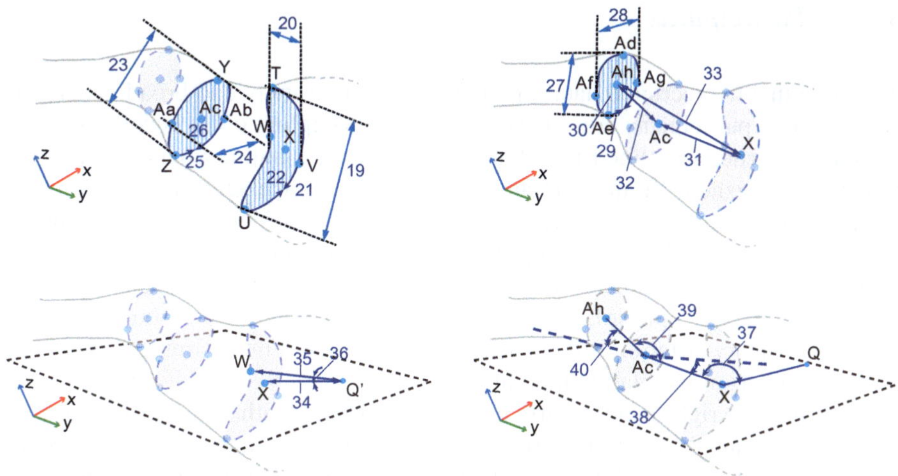

Fig. 3.2 Variables of characterization for the external acoustic meatus: variables (19–40) include entrance length (19), entrance width (20), entrance circumference (21), entrance area (22), first bend length (23), first bend width (24), first bend circumference (25), first bend area (26), second bend length (27), second bend width (28), second bend circumference (29), second bend area (30), section I depth (31), Section II depth from first bend (32), Section II depth from entrance (33), entrance depth from geometric centre to medial concha (34), entrance depth at posterior from medial concha (35), azimuth angle of geometric centre of entrance (36), azimuth angle of Section I (37) , elevation angle of Section I (38), azimuth angle of Section II (39), elevation angle of Section II (40); landmarks (T-Ah) include superior entrance (T), inferior entrance (U), anterior entrance (V), posterior entrance (W), geometric centre of entrance (X), superior first bend (Y), inferior first bend (Z), anterior first bend (Aa), posterior first bend (Ab), geometric centre of first bend (Ac), superior second bend (Ad), inferior second bend (Ae), anterior second bend (Af), posterior second bend (Ag), geometric centre of second bend (Ah), medial concha (projection) (Q')

within the cranial region, providing a basis for designing the overall morphology of in-ear components.

3.2 Shape-Structure Characterization Methods

The characterization integrates the auricle structure and the external acoustic meatus (EAM) structure described in Sect. 2.3, together with 40 characterization variables, to establish a methodological framework for the external ear. This framework encompasses participant recruitment, the definition of anatomical landmarks, the construction of reference orientations, axes, and planes, the specification of measurement variables, and the procedures for data collection, processing, and statistical modelling. Within this framework, the Chinese population is characterized in terms of growth trajectories, bilateral asymmetries, and sex-related differences. The foundation for subsequent chapters is also provided through the presentation of characterization outcomes, thresholds of wearing comfort, and perception limits.

3.2.1 Participants

To ensure that the recruited data capture population diversity and broad geographical coverage, participants are selected from eight sampling cities according to the recruitment criteria outlined in Table 2.2: Shenyang, Tianjin, Xuzhou, Shanghai, Fuzhou, Guangzhou, Xi'an, and Lanzhou. Data collection extends over a 28-month period. Representativeness is achieved primarily by considering the geographical distribution of population density in China, taking into account the National Population Census and the marked population density differences across the Hu Huanyong Line. Except for Lanzhou, all sampling cities are located southeast of the line. A secondary criterion is that each city functions as a major transportation hub with high levels of population mobility.

In total, 1120 right-handed participants are recruited, with native provinces covering more than 88% of the provincial-level administrative regions in China. Consistent with the principles described in Sect. 2.3, no participants are included from Xinjiang, Hong Kong, Macao, or Taiwan. Zhao et al. [22] indicate that the minimum age at which external ear maturity is achieved in the Chinese population is 14 years; accordingly, the recruitment age is set at 15 years. A pilot study conducted from 2016 to 2017 using equidistant age grouping [11] shows that among 700 participants, fewer than 10 are aged 80 years or older. On this basis, the maximum recruitment age is restricted to 79 years. Following the 10-year grouping method (10-year intervals per group), participants are divided into seven age groups: teens (15–19), 20 s (20–29), 30 s (30–39), 40 s (40–49), 50 s (50–59), 60 s (60–69), and 70 s (70–79).

The data sampling process is considered complete once the following conditions are met: (1) equal numbers of participants in each age group, balanced by sex; (2) three-dimensional ear models of both ears contain complete data through the second bend of the EAM; and (3) within each age group, at least two participants (one male and one female) originate from the same provincial-level administrative unit.

Ultimately, 560 males and 560 females are recruited, with a minimum of 12 participants (6 males and 6 females) included for each age group in each sampling city. Exclusion criteria related to auricular health, piercing history, family relations, and kinship are summarized in Table 2.2. Participant recruitment locations and the distribution of native provinces are illustrated in Fig. 3.3.

3.2.2 Anatomical Landmarks

As illustrated in Fig. 3.4, a total of 34 anatomical landmarks is defined to support the measurement variables. Of these, 19 landmarks (A-S) relate to the auricle. Points G (otobasion superius), H (otobasion inferius), K (superior cavum concha), M (tragion), Q (medial concha), and R (otobasion posterius) are adapted from the frameworks of Fan et al. [11], Lee et al. [7], and Roebuck [8], while the remaining auricle landmarks draw on Purkait [13], Zhao et al. [22], and Farkas [17]. In

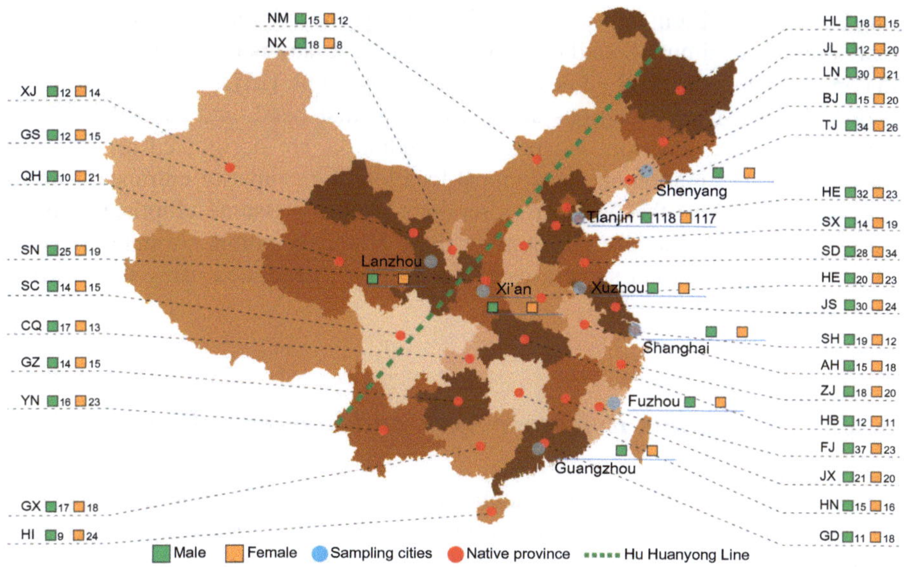

Fig. 3.3 Distribution of participants across sampling locations and native provinces in the shape-structure characterization

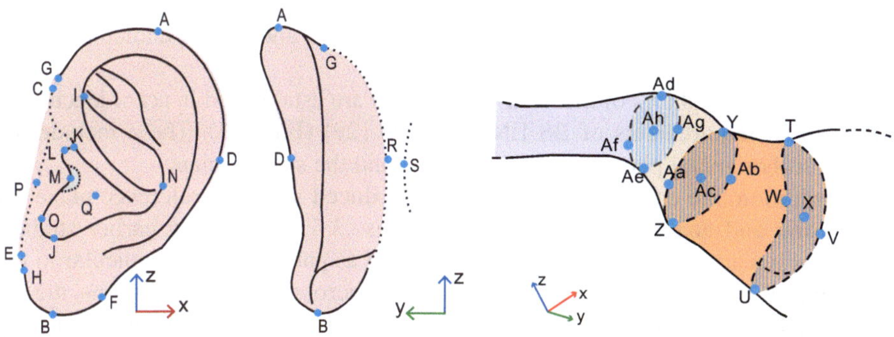

Fig. 3.4 Anatomical landmarks: superaurale (A), subaurale (B), preaurale (C), postaurale (D), anterior lobule (E), posterior lobule (F), otobasion superius (G), otobasion inferius (H), anterior cymba concha (I), incisura intertragica inferior (J), superior cavum concha (K), incisura anterior auris posterior (L), tragion (M), posterior concha (N), lowest point on the lower border of tragus (O), cardinal point of tragus (P), medial concha (Q), otobasion posterius (R), anterior mastoid (S); superior entrance (T), inferior entrance (U), anterior entrance (V), posterior entrance (W), geometric centre of entrance (X), superior first bend (Y), inferior first bend (Z), anterior first bend (Aa), posterior first bend (Ab), geometric centre of first bend (Ac), superior second bend (Ad), inferior second bend (Ae), anterior second bend (Af), posterior second bend (Ag), geometric centre of second bend (Ah)

addition, 15 EAM-related landmarks (X-Ah) are newly defined. Among them, X (geometric centre of the entrance), Ac (geometric centre of the first bend), and Ah (geometric centre of the second bend), are generated using the "find-circle-centre" function in RapidForm XOR3 (INUS Technology, Inc., Korea). The other 12 landmarks serve as azimuth boundary points (orientation landmarks), named according to conventions described by Lee et al. [18] and Fan et al. [25].

For the EAM cross-sections at the entrance, first bend, and second bend, the superior, inferior, anterior, and posterior landmarks are designated according to their specific orientation and anatomical location (e.g., superior entrance, anterior first bend). Because the EAM cross-sections typically assume elliptical or ovoid forms [12, 24], the precise positions of these boundary points are determined by identifying the long and short axes of the sections, generated using the "extract-axis" function in RapidForm.

3.2.3 Orientation, Reference Axes, and Reference Planes

The orientation definitions of auricular anatomy and engineering anthropometry follow the auricular axis construction method (Fig. 2.9a, AIA inclination orientation). The angular difference between the orientation defined by the constructed auricular axis and that defined by the head-face system based on the Frankfurt horizontal plane represents the auricular inclination angle (AIA). The reference axes include the long axis of the auricle and the transverse axis of the auricle at the tragion. The long axis passes through the superaurale and subaurale, while the transverse axis at the tragion passes through the tragion and is perpendicular to the long axis.

The reference planes of the head-face system are established in accordance with the general requirements of BS EN 960:2006 [29] (Fig. 3.5). These include the Frankfurt horizontal plane, the coronal plane, and the sagittal plane.

In addition, the ear horizontal plane is introduced as a new reference plane for measuring and analysing EAM variables (Figs. 3.5 and 3.6). In earlier studies, auricular linear variables were generally defined as parallel or perpendicular to the long axis of the auricle. To maintain consistency across variable definitions, the ear horizontal plane is specifically defined as parallel to the transverse axis of the

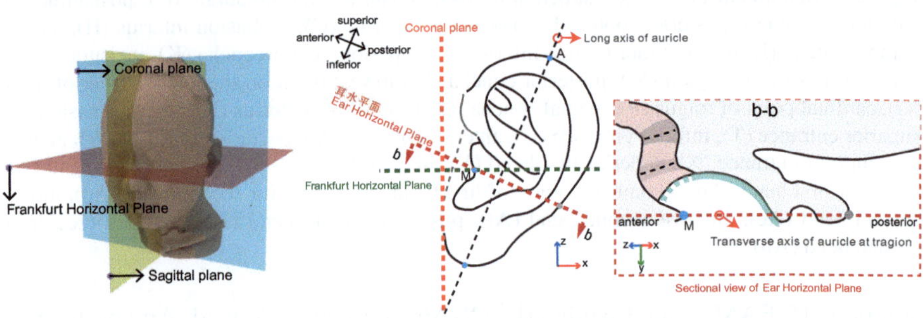

Fig. 3.5 Reference planes and axes: landmarks include superaurale (A), subaurale (B), tragion (M)

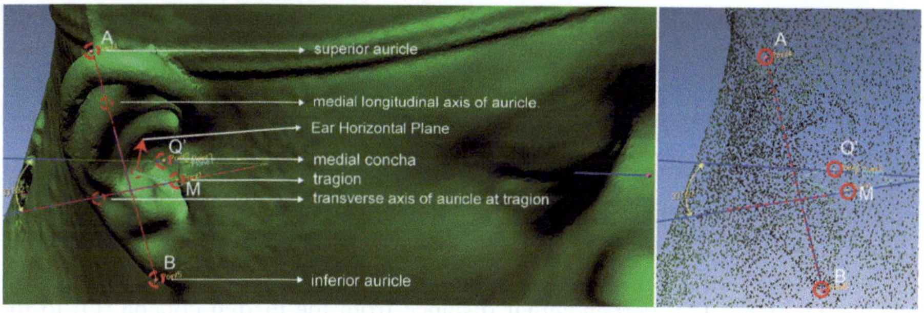

Fig. 3.6 Relative position of Ear Horizontal Plane on the head-face model: landmarks include superaurale (A), subaurale (B), tragion (M), medial concha (projection) (Q')

auricle at the tragion and perpendicular to both the long axis of the auricle and the sagittal plane of the head-face system.

3.2.4 Variable Definitions

The 40 measurement variables used are defined as follows:

1. AL (Variable 1): Distance between the superaurale (A) and the subaurale (B), parallel to the auricular long axis.
2. AWP (Variable 2): Distance between the preaurale (C) and the postaurale (D), perpendicular to the auricular long axis.
3. AWI (Variable 3): Distance between the posterior incisura anterior auris (L) and the posterior auricular margin, parallel to the auricular long axis.
4. LL (Variable 4): Distance between the inferior incisura intertragica (J) and the subaurale (B), perpendicular to the Frankfurt horizontal plane.
5. LW (Variable 5): Distance between the anterior lobule (E) and the posterior lobule (F), perpendicular to the auricular long axis.
6. EML (Variable 6): Distance between the otobasion superius (G) and the otobasion inferius (H).
7. AIA (Variable 7): Angle between the auricular long axis and the coronal plane.
8. CL (Variable 8): Distance between the anterior cymba concha (I) and the inferior incisura intertragica (J), parallel to the auricular long axis.
9. CW (Variable 9): Distance between the posterior incisura anterior auris (L) and the posterior concha (N), perpendicular to the auricular long axis.
10. CCL (Variable 10): Distance between the inferior incisura intertragica (J) and the superior cavum concha (K), parallel to the auricular long axis.
11. CCW (Variable 11): Distance between the tragion (M) and the posterior concha (N), perpendicular to the auricular long axis.

12. TL (Variable 12): Distance between the posterior incisura anterior auris (L) and the lowest point on the lower border of the tragus (O), parallel to the auricular long axis.
13. TH (Variable 13): Distance between the tragion (M) and the cardinal point of the tragus (P), perpendicular to the auricular long axis.
14. EPS (Variable 14): Distance from the superaurale (A) to the head, perpendicular to the sagittal plane.
15. EPT (Variable 15): Distance from the postaurale (D) to the head, perpendicular to the sagittal plane.
16. CD (Variable 16): Perpendicular distance from the medial concha (Q) to the line connecting the tragion (M) and the posterior concha (N), parallel to the ear horizontal plane.
17. CAP (Variable 17): Concho-mastoid angle at the postaurale (AWP, Variable 2), parallel to the ear horizontal plane.
18. CAO (Variable 18): Concho-mastoid angle at the otobasion posterius (AWI, Variable 3), with the vertex at R, parallel to the ear horizontal plane.
19. EL (Variable 19): Distance between the superior entrance (T) and the inferior entrance (U).
20. EW (Variable 20): Distance between the anterior entrance (V) and the posterior entrance (W).
21. EC (Variable 21): Circumference of the entrance of the EAM.
22. EA (Variable 22): Area of the entrance cross-section.
23. FBL (Variable 23): Distance between the superior first bend (Y) and the inferior first bend (Z).
24. FBW (Variable 24): Distance between the anterior first bend (Aa) and the posterior first bend (Ab).
25. FBC (Variable 25): Circumference of the first bend of the EAM.
26. FBA (Variable 26): Area of the first bend of the EAM.
27. SBL (Variable 27): Distance between the superior second bend (Ad) and the inferior second bend (Ae).
28. SBW (Variable 28): Distance between the anterior second bend (Af) and the posterior second bend (Ag).
29. SBC (Variable 29): Circumference of the second bend of the EAM.
30. SBA (Variable 30): Area of the second bend of the EAM.
31. SD-I (Variable 31): Distance between the geometric centre of the entrance (X) and the geometric centre of the first bend (Ac).
32. SD-II-F (Variable 32): Distance between the geometric centre of the first bend (Ac) and the geometric centre of the second bend (Ah).
33. SD-II-E (Variable 33): Distance between the geometric centre of the entrance (X) and the geometric centre of the second bend (Ah).
34. ED-C (Variable 34): Distance between the medial concha (Q) and the posterior entrance (W), parallel to the ear horizontal plane.
35. ED-P (Variable 35): Distance between the medial concha (Q) and the geometric centre of the entrance (X), parallel to the ear horizontal plane.

36. ACE (Variable 36): Angle between ED-C (Variable 34) and ED-P (Variable 35), with the vertex at the medial concha (Q), parallel to the ear horizontal plane.
37. AS-I (Variable 37): Angle between SD-I (Variable 31) and ED-P (Variable 35), with the vertex at the geometric centre of the entrance (X).
38. ES-I (Variable 38): Angle between SD-I (Variable 31) and its projection on the ear horizontal plane, with the vertex at the geometric centre of the first bend (Ac).
39. AS-II (Variable 39): Angle between SD-I (Variable 31) and SD-II-F (Variable 32), with the vertex at the geometric centre of the first bend (Ac).
40. ES-II (Variable 40): Angle between SD-II-F (Variable 32) and its projection on the ear horizontal plane, with the vertex at the geometric centre of the first bend (Ac).

3.2.5 Data Collection and Processing

Three-dimensional data of the head-face region and both ears are acquired indirectly using non-contact techniques. The 3D models of the head-face and auricle are obtained through 3D scanning (EinScan-Pro+ Handheld HD Scanner, SHINING 3D Inc., China; accuracy: 0.05 mm; spatial point interval: 0.25 mm). The 3D model of the external acoustic meatus (EAM) is generated by combining impression molding with 3D scanning (ABR impression materials, SOUNDLINK, Inc., China). As illustrated in Fig. 3.7, the workflow for external ear 3D data acquisition consists of the following steps:

1. Scanning the head and EAM impression mold using a handheld HD scanner;
2. Performing hole-filling and smoothing of the 3D models (point clouds) with the EinScan-Pro Series v2.6.0.9 auxiliary software;
3. Extracting the auricle model from the head-face model and merging it with the EAM mold using RapidForm;

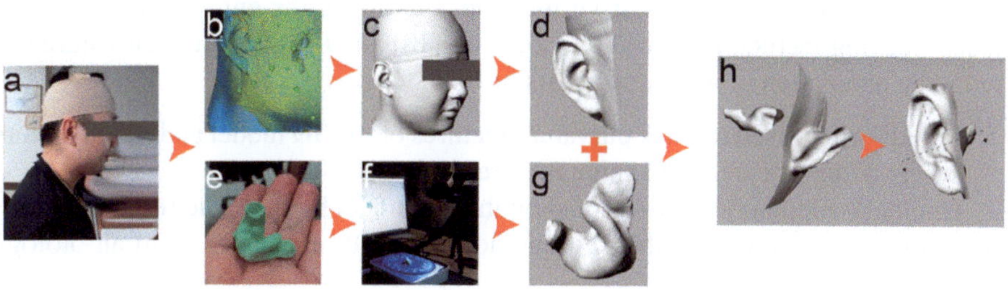

Fig. 3.7 Steps involved in the process of casting and scanning the external ear (EAM: external acoustic meatus): (**a**) recruiting subjects, (**b**) scanning head and face, (**c**) filling, smoothing and meshing using the point cloud model of the head and face, (**d**) extracting 3D model of the auricle, (**e**) casting of the concha and EAM, (**f**) a scanning impression prototype of the concha and EAM, (**g**) filling, smoothing and meshing point cloud model of concha and EAM, (**h**) merging of models to create a 3D model of the external ear

4. Meshing the merged external ear 3D model (point cloud).

Anatomical and measurement landmarks are marked on the external ear 3D model using RapidForm. For the characterization variables, measurement values of the linear, angular, and circumference types are obtained by exporting the corresponding landmark coordinates into MATLAB and applying computational procedures. Measurement values of the area variables are extracted directly from the 3D model using the "section extraction" function in RapidForm.

The measurement of linear variables is based on the Euclidean distance formula in three-dimensional space (Eq. (3.1)). Angular variables are derived using the vector angle formula in three-dimensional space (Eq. (3.2)). Circumference variables are computed from the cross-sectional circumference formula in three-dimensional space (Eq. (3.3)).

$$Dist(a,b) = \sqrt{(x_a - x_b)^2 + (y_a - y_b)^2 + (z_a - z_b)^2} \tag{3.1}$$

$$Arccos\theta = Arc\left(\frac{p \cdot q}{|p||q|}\right) = Arc\left(\frac{x_p x_q + y_p y_q + z_p z_q}{\sqrt{x_p^2 + y_p^2 + z_p^2} + \sqrt{x_q^2 + y_q^2 + z_q^2}}\right) \tag{3.2}$$

$$C_{cross-section} = \sqrt{(x_n - x_1)^2 + (y_n - y_1)^2 + (z_n - z_1)^2} + \sum_{i=1}^{n} \sqrt{(x_{i+1} - x_i)^2 + (y_{i+1} - y_i)^2 + (z_{i+1} - z_i)^2} \tag{3.3}$$

where, a and b are points in three-dimensional space, with $a = (x_a, y_a, z_a)$ and $=(x_b, y_b, z_b)$; $\boldsymbol{p}$ and $\boldsymbol{q}$ are vectors in three-dimensional space, with $\boldsymbol{p} = [x_p, y_p, z_p]^T$ and $\boldsymbol{q} = [x_q, y_q, z_q]^T$; $\boldsymbol{n}$ denotes the number of landmarks, and (x_i, y_i, z_i) represents the coordinates of the i-th point.

To minimize inter-observer error, landmarking of 3D models is carried out by two designated researchers. Researcher A performs landmark annotation, while Researcher B is responsible for verification. Following the engineering software-aided tracking method proposed by Lee et al. [18] and the maximum curvature criteria, the annotation procedure consists of four steps:

1. RapidForm is used to track curvature features of the 3D model and retain feature points within the intended annotation region;
2. landmark annotation is performed on the 3D model by the same researcher;
3. landmark positions are checked and adjusted with assistance from an otology specialist;
4. Intra-measure reliability of landmarks is evaluated using the methods described Lee et al. [18].

Evaluation of intra-measure reliability is conducted as follows. 3D models are randomly selected from the dataset, and each landmark is annotated three times by the same researcher, with at least 24 hours between operations on the same site. When the standard deviation (SD) of repeated measurements is less than 2 mm and the

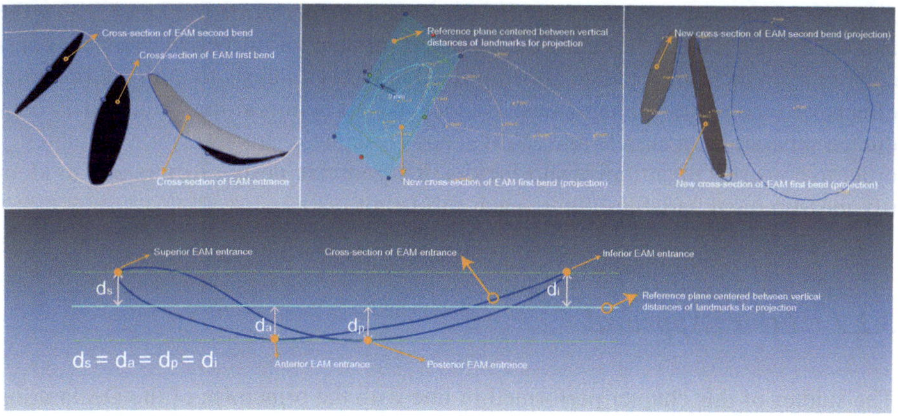

Fig. 3.8 Creation of centred reference plane

coefficient of variation (CV) is less than 5%, the original annotation is regarded as reliable.

In approximately 6% of participants (all aged 64 years or older), contour curves of the EAM entrance and the two bends appear as spatial curves, making it impossible to generate orientation landmarks directly. In these cases, the data processing method involves selecting the point located centrally at the orientation inflection among software-recommended points as the boundary landmark. A centred reference plane of the sectional landmarks is then created and used as the projection plane for subsequent data processing [30, 31] (Fig. 3.8).

3.2.6 Statistical Analysis

Statistical analyses are carried out in MATLAB under a 95% confidence level ($\alpha = 0.05$). Results are regarded as statistically significant when $p < 0.05$. The Shapiro-Wilk test (S-W test) is used to evaluate the normality of the sampled data.

In the difference analysis, one-way analysis of variance (One-way ANOVA) is applied to assess the overall influence of age groups on the 40 shape-structure characterization variables, that is, to determine whether distributions differ significantly across age groups. Tukey's honest significant difference test (Tukey HSD test) is employed for post hoc multiple comparisons among groups. The Pearson correlation coefficient (r) is used to examine correlations between characterization variables and participants' age, height, weight, and body mass index (BMI). Paired-sample t-tests and independent-sample t-tests are adopted to analyse bilateral and sex-related differences in the characterization variables. When independent-sample t-tests reveal significant differences, height and weight serve as covariates to adjust variable differences, and Bonferroni correction is applied for further post hoc comparisons.

In the factor analysis, KMO and Bartlett's tests are performed. When KMO > 0.6 and Bartlett's $p < 0.05$, the data are considered suitable for further analysis. Z-scores derived from the sampled data are employed in factor analysis to mitigate potential effects of inconsistent dimensions. Factor analysis with varimax rotation is conducted separately on auricle and EAM shape-structure characterization variables in order to construct quantitative sub-factors of ergonomic design factors and to identify measurement variables within each sub-factor.

3.2.7 *Ethics Approval*

The studies of this chapter performed in line with the principles of the Declaration of Helsinki. Informed consent was obtained from all participants. The studies of this chapter were approved by the Ethics Committee of Northwestern Polytechnic University (No. IRB No. 201502024, No. 202002024).

3.3 Difference Analysis

In the characterization stage, valid data are obtained from 1120 participants. A total of 2240 external ears is included in the 3D measurement dataset for analysis. The demographic characteristics of the participants are summarized in Table 3.3, with further details provided in Table B.1.

3.3.1 *Descriptive Statistics and Correlation Analysis*

Descriptive Statistics and Age Group Differences

As shown in Table 3.4, among the auricular linear variables, the mean value of AL is the largest (66.12 ± 5.60 mm), whereas the mean value of TH is the smallest (8.37 ± 1.24 mm). The mean width-to-length ratios are 54.04% for the superior contour ($\frac{AWP}{AL}$ %), 42.68% for the inferior contour ($\frac{AWI}{AL}$ %), and 85.34% for the auricular base-to-length ratio ($\frac{EML}{AL}$ %). These results suggest that the auricle as a whole can be enclosed within a rectangular boundary box. For the concha and cavum concha, mean values of length variables exceed those of width variables, with width-to-length ratios of 66.60% ($\frac{CW}{CL}$ %) and 83.49% ($\frac{CCW}{CCL}$ %), respectively. The mean height-to-length ratio of the tragus is approximately 45.86% ($\frac{TH}{TL}$

Table 3.3 Demographics of participants

Gender	N	Age (years)	Height (m)	Weight (kg)	BMI (kg/m^2)
		Mean ± SD	Mean ± SD	Mean ± SD	Mean ± SD
Male	560	44.96 ± 19.39	1.72 ± 0.06	71.41 ± 9.58	24.20 ± 3.03
Female	560	44.71 ± 19.41	1.63 ± 0.06	56.38 ± 8.74	21.23 ± 3.33

Table 3.4 Descriptive statistics of characterization variables for the auricle

		Ear, N = 2240				ANOVA	
No.	Variable	Mean	SD	Max	Min	df (age group)	p
1	AL (mm)	66.12	5.60	48.05	79.05	6	< 0.001
2	AWP (mm)	35.74	3.50	26.96	48.96	6	< 0.001
3	AWI (mm)	28.22	2.60	19.68	40.32	6	< 0.001
4	LL (mm)	19.95	3.02	12.33	28.64	6	< 0.001
5	LW (mm)	19.15	2.45	10.05	28.33	6	< 0.001
6	EML (mm)	56.43	5.51	38.52	70.00	6	< 0.001
7	AIA (°)	19.03	4.47	11.21	35.10	6	< 0.001
8	CL (mm)	29.46	3.13	16.95	39.78	6	< 0.001
9	CW (mm)	19.62	2.20	12.14	28.54	6	< 0.001
10	CCL (mm)	18.41	1.88	12.05	23.93	6	< 0.001
11	CCW (mm)	15.37	2.21	8.59	24.97	6	< 0.001
12	TL (mm)	18.25	1.87	12.00	23.84	6	< 0.001
13	TH (mm)	8.37	1.24	4.35	12.50	6	< 0.001
14	EPS (mm)	16.63	2.44	10.05	24.23	6	< 0.001
15	EPT (mm)	23.84	3.32	16.15	55.84	6	< 0.001
16	CD (mm)	16.30	2.39	9.11	23.96	6	< 0.001
17	CAP (°)	41.91	5.38	24.03	55.47	6	< 0.001
18	CAO (°)	57.36	12.23	31.23	90.01	6	< 0.001

%), indicating that both the concha and the tragus can also be approximately enclosed within rectangular boundary boxes. Relative to the coronal plane, the auricle on average inclines posteriorly toward the occipital bone by 19.03 ± 4.47° (AIA). Among the angular variables, CAO records the largest mean value (57.36 ± 12.23°), whereas AIA is the smallest. The superior protrusion variables (EPS and CAP) present smaller mean values than the posterior protrusion variables (EPT and CAO), with superior-to-posterior ratios of 69.76% ($\frac{EPS}{EPT}$ %) for linear distance and 73.06% ($\frac{CAP}{CAO}$ %) for angular distance. These findings indicate that auricular protrusion follows a consistent pattern of being smaller superiorly and larger posteriorly.

As shown in Table 3.5, for the external acoustic meatus (EAM), the entrance variables (EL, EW, EC, and EA) display larger mean values compared with those of the first bend (FBL, FBW, FBC, and FBA) and the second bend (SBL, SBW,

Table 3.5 Descriptive statistics of characterization variables for the external acoustic meatus

		Ear, N = 2240				ANOVA	
No.	Variable	Mean	SD	Max	Min	df (age group)	p
19	EL (mm)	15.35	1.34	9.59	17.99	6	<0.001
20	EW (mm)	8.60	0.88	6.00	11.91	6	<0.001
21	EC (mm)	40.10	3.25	29.01	52.13	6	<0.001
22	EA (mm^2)	104.80	15.63	55.79	155.31	6	<0.001
23	FBL (mm)	9.68	1.19	6.00	12.77	6	<0.001
24	FBW (mm)	7.70	0.94	5.00	10.00	6	<0.001
25	FBC (mm)	27.98	2.75	19.56	34.74	6	<0.001
26	FBA (mm^2)	58.52	11.03	28.43	90.91	6	<0.001
27	SBL (mm)	7.91	0.77	4.21	9.98	6	<0.001
28	SBW (mm)	7.25	0.87	3.46	9.98	6	<0.001
29	SBC (mm)	24.16	2.09	16.45	35.48	6	<0.001
30	SBA (mm^2)	45.33	7.26	21.80	68.79	6	<0.001
31	SD-I (mm)	7.51	1.26	4.08	11.91	6	<0.001
32	SD-II-F (mm)	7.28	1.36	5.00	13.86	6	<0.001
33	SD-II-E (mm)	13.54	2.10	9.05	21.78	6	<0.001
34	ED-C (mm)	14.10	1.48	10.00	18.87	6	<0.001
35	ED-P (mm)	10.96	1.36	6.00	14.69	6	<0.001
36	ACE (°)	24.82	3.06	16.02	33.90	6	<0.001
37	AS-I (°)	108.33	6.35	88.14	119.94	6	<0.001
38	ES-I (°)	15.26	5.42	1.37	39.86	6	<0.001
39	AS-II (°)	133.95	8.82	101.50	158.26	6	<0.001
40	ES-II (°)	30.15	6.51	15.98	50.74	6	<0.001

SBC, and SBA). The mean values of the second bend are the smallest among the three cross-sections. For entrance localization variables, the mean of ED-C exceeds that of ED-P, and the angular range of ACE is 16.02°-33.90°, suggesting that the EAM entrance tends to deviate anteriorly relative to the coronal plane. For canal section depth variables (SD-I, SD-II-F, and SD-II-E), the values form a canal-section triangle, with the relationship SD-II-F < SD-I < SD-II-E. For canal section localization angles, elevation angles follow the pattern ES-I < ES-II, while mean values of azimuth angles (AS-I and AS-II) exceed 100°. These results indicate that within the skull, canal sections progressively deviate superiorly and medially toward the mid-sagittal plane. The minimum value of ES-I is 1.37°, showing that in some participants, the first canal section is nearly parallel to the ear horizontal plane. Percentile results are presented in Tables B.2 and B.3. As summarized in Tables 3.4 and 3.5, all 40 characterization variables exhibit significant age-group differences ($p < 0.001$). Subsequent subsections examine age-variability and age-variation phenomena in detail on the basis of these groupwise distributions.

Table 3.6 Correlation analysis of characterization variable for the auricle

No.	Variable	Age		Height		Weight		BMI	
		Pearson correlation	Sig. (2-tailed)	Pearson correlation	Sig. (2-tailed)	Pearson correlation	Sig. (2-tailed)	Pearson correlation	Sig. (2-tailed)
1	AL	0.79[a]	<0.001	0.10[a]	<0.001	0.18[a]	<0.001	0.28[a]	<0.001
2	AWP	0.62[a]	<0.001	0.08[a]	0.01	0.15[a]	<0.001	0.23[a]	<0.001
3	AWI	0.49[a]	<0.001	0.07[b]	0.03	0.08[a]	0.01	0.14[a]	<0.001
4	LL	0.58[a]	<0.001	0.07[b]	0.03	0.13[a]	<0.001	0.20[a]	<0.001
5	LW	0.38[a]	<0.001	0.10[a]	0.002	−0.02	0.56	0.04	0.22
6	EML	0.65[a]	<0.001	−0.03	0.27	0.19[a]	<0.001	0.25[a]	<0.001
7	AIA	0.04	0.14	0.04	0.18	0.06	0.06	0.04	0.12
8	CL	0.50[a]	<0.001	−0.05	0.14	0.13[a]	<0.001	0.19[a]	<0.001
9	CW	0.33[a]	<0.001	0.002	0.94	0.11[a]	<0.001	0.14[a]	<0.001
10	CCL	0.21[a]	<0.001	0.02	0.48	0.10[a]	0.001	0.11[a]	<0.001
11	CCW	0.32[a]	<0.001	−0.01	0.76	0.09[a]	0.003	0.12[a]	<0.001
12	TL	0.21[a]	<0.001	0.04	0.21	0.11[a]	<0.001	0.12[a]	<0.001
13	TH	0.002	0.95	0.08[b]	0.01	0.10[a]	<0.001	0.08[a]	0.01
14	EPS	0.05[b]	<0.001	0.12[a]	<0.001	0.14[a]	<0.001	0.11[a]	<0.001
15	EPT	0.06[b]	<0.001	0.13[a]	<0.001	0.11[a]	<0.001	0.06[b]	0.04
16	CD	0.03	0.35	0.14[a]	<0.001	0.11[a]	<0.001	0.05	0.08
17	CAP	−0.28[a]	<0.001	0.23[a]	<0.001	0.13[a]	<0.001	0.02	0.50
18	CAO	−0.14[a]	<0.001	0.18[a]	<0.001	0.10[a]	0.001	0.02	0.52

[a] Correlation is significant at the 0.01 level (2-tailed)
[b] Correlation is significant at the 0.05 level (2-tailed)

Correlation Analysis

The correlations of measurement variables with age, height, weight, and BMI are summarized in Tables 3.6 and 3.7. Among the auricular variables, AL shows the strongest correlation with age ($r \approx 0.79$, $p < 0.001$). By contrast, AIA, TH, and CD exhibit no significant correlations with age, indicating that although these variables differ significantly across age groups, they do not follow consistent trends associated with age progression. In particular, AIA displays no significant correlation with age, height, weight, or BMI. Thirteen auricular variables demonstrate significant correlations with BMI, suggesting that participants with higher BMI tend to present larger auricular dimensions, a tendency primarily attributable to increased thickness of the helix, tragus, and lobule skin. This finding indirectly echoes the literal meaning of the Chinese idiom "big ears on a plump head."

For EAM variables, EL records the strongest correlation with age ($r \approx 0.67$, $p < 0.001$). No significant correlations with height are observed for SBW, AS-I, ES-I, AS-II, and ES-II, indicating that these variables do not exhibit ordered variation across different statues. By contrast, SD-I, SD-II-F, SD-II-E, ED-C, ED-P, and ACE are significantly and negatively correlated with height. This pattern suggests that participants with greater stature are more likely to exhibit a shorter EAM length,

Table 3.7 Correlation analysis of characterization variable for the external acoustic meatus

No.	Variable	Age		Height		Weight		BMI	
		Pearson correlation	Sig. (2-tailed)	Pearson correlation	Sig. (2-tailed)	Pearson correlation	Sig. (2-tailed)	Pearson correlation	Sig. (2-tailed)
19	EL	0.67[a]	<0.001	0.12[a]	<0.001	0.30[a]	<0.001	0.35[a]	<0.001
20	EW	0.18[a]	<0.001	0.13[a]	<0.001	0.20[a]	<0.001	0.17[a]	<0.001
21	EC	0.56[a]	<0.001	0.05[b]	0.04	0.30[a]	<0.001	0.33[a]	<0.001
22	EA	0.53[a]	<0.001	0.07[a]	0.01	0.29[a]	<0.001	0.32[a]	<0.001
23	FBL	0.55[a]	<0.001	0.09[a]	0.001	0.29[a]	<0.001	0.30[a]	<0.001
24	FBW	0.20[a]	<0.001	0.11[a]	<0.001	0.21[a]	<0.001	0.19[a]	<0.001
25	FBC	0.50[a]	<0.001	0.12[a]	<0.001	0.31[a]	<0.001	0.31[a]	<0.001
26	FBA	0.45[a]	<0.001	0.13[a]	<0.001	0.31[a]	<0.001	0.30[a]	<0.001
27	SBL	0.12[a]	<0.001	0.17[a]	<0.001	0.20[a]	<0.001	0.15[a]	<0.001
28	SBW	0.37[a]	<0.001	0.02	0.46	0.17[a]	<0.001	0.19[a]	<0.001
29	SBC	0.30[a]	<0.001	0.12[a]	<0.001	0.23[a]	<0.001	0.20[a]	<0.001
30	SBA	0.33[a]	<0.001	0.10[a]	<0.001	0.22[a]	<0.001	0.21[a]	<0.001
31	SD-I	0.57[a]	<0.001	−0.12[a]	<0.001	0.25[a]	<0.001	0.30[a]	<0.001
32	SD-II-F	0.29[a]	<0.001	−0.11[b]	<0.001	0.06[b]	0.02	0.14[a]	<0.001
33	SD-II-E	0.56[a]	<0.001	−0.05[b]	0.04	0.20[a]	<0.001	0.28[a]	<0.001
34	ED-C	0.55[a]	<0.001	−0.09[a]	0.001	0.10[a]	<0.001	0.18[a]	<0.001
35	ED-P	0.60[a]	<0.001	−0.10[a]	<0.001	0.13[a]	<0.001	0.22[a]	<0.001
36	ACE	0.61[a]	<0.001	−0.10[a]	<0.001	0.17[a]	<0.001	0.27[a]	<0.001
37	AS-I	0.34[a]	<0.001	0.03	0.23	0.13[a]	<0.001	0.15[a]	<0.001
38	ES-I	0.39[a]	<0.001	−0.05	0.10	0.09[a]	0.001	0.14[a]	<0.001
39	AS-II	0.20[a]	<0.001	0.002	0.95	0.08[a]	0.005	0.10[a]	<0.001
40	ES-II	0.27[a]	<0.001	0.03	0.33	0.10[a]	<0.001	0.11[a]	<0.001

[a] Correlation is significant at the 0.01 level (2-tailed)
[b] Correlation is significant at the 0.05 level (2-tailed)

a posterior entrance located closer to the medial concha, and a greater degree of entrance deviation toward the sagittal plane.

3.3.2 *Persistent Age-Growth and Phasic Age-Growth Phenomenon*

The analyses are based on cross-sectional data obtained from different age groups. Figure 3.9 illustrates representative auricular photographs, captured at equal distances, together with three-dimensional scans of the concha and partial external acoustic meatus (EAM, obtained after hair removal and casting) that supplement the pilot study. The participants shown represent age groups of 20, 30, 40, 50, 60, 70, 80, and 90 years.

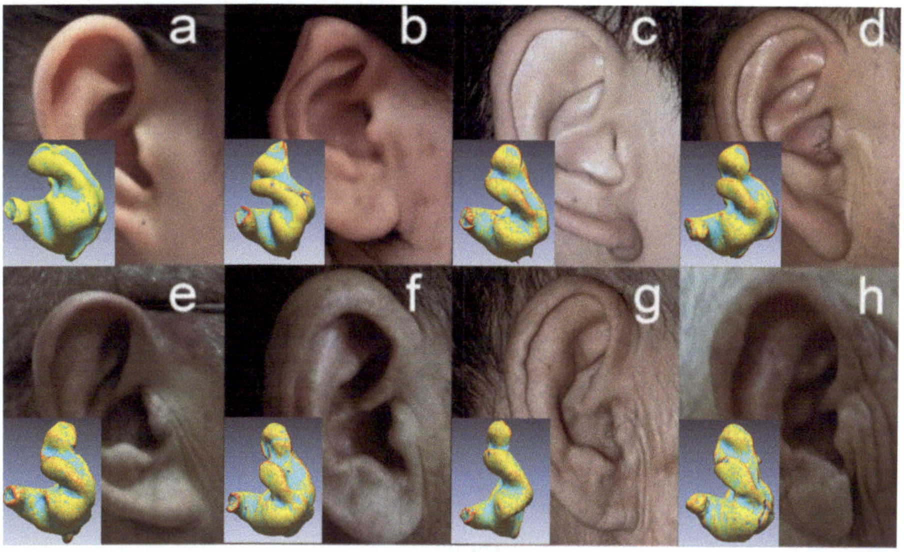

Fig. 3.9 Samples of the auricle, concha and (part of) external acoustic meatus: (**a**) 20 year-old; (**b**) 30 year-old; (**c**) 40 year-old; (**d**) 50 year-old; (**e**) 60 year-old; (**f**) 70 year-old; (**g**) 80 year-old; (**h**) 90 year-old

Visually, the age-related changes of the external ear reveal clear trends. The overall size of the auricle enlarges with age, the skin thickens and develops more wrinkles, and the skin surface of the concha and the entrance of the EAM becomes progressively rougher. Tables 3.8 and 3.9 present the multiple comparison results of 40 measurement variables across age groups, while detailed statistical values are reported in Tables B.4 and B.5. These findings form the basis for distinguishing between two categories of age-related variation: persistent age-growth variables, which continue to increase even in older age groups, and phasic age-growth variables, which exhibit distinct stages of increase, stabilization, and decline.

For the purposes of this book, these categories are defined as follows:

1. Persistent age-growth variables refer to quantitative morphological variables of the external ear that show a consistent and monotonic increase with advancing age, including in older age groups, without evidence of plateau or decline. Such variables represent structural dimensions that enlarge throughout the lifespan, reflecting a persistent pattern of age-related growth.
2. Phasic age-growth variables refer to quantitative morphological variables of the external ear that display a non-monotonic trajectory of change. These are characterized by an initial phase of increase during early and middle age, followed by a plateau and an eventual decline in older age. Such variables represent age-dependent structural dynamics that undergo stage-specific transitions rather than continuous enlargement.

Table 3.8 Multiple comparisons of characterization variables for the auricle (by age groups)

		Multiple comparisons (age group)						
No.	Variable	a	b	c	d	e	f	g
1	AL	10s	20s	30s	40s	50s	60s	70s
2	AWP	10s	20s	30s, 40s	50s, 60s	70s		
3	AWI	10s	20s	30s, 40s	50s	60s	70s	
4	LL	10s	20s, 30s, 40s	50s	60s	70s		
5	LW	10s, 20s, 30s	40s	50s, 60s	70s			
6	EML	10s	20s	30s, 40s	50s, 60s	70s		
7	AIA	10s, 50s	20s, 40s, 70s	20s, 30s	60s			
8	CL	10s	20s, 30s	40s	50s, 60s	70s		
9	CW	10s	20s, 30s	40s	50s, 60s, 70s			
10	CCL	10s	20s, 30s	40s, 50s, 60s, 70s				
11	CCW	10s	20s, 30s	40s	50s	60s, 70s		
12	TL	10s	20s, 30s	40s, 50s, 60s, 70s				
13	TH	10s, 70s	20s, 60s, 70s	20s, 30s, 40s, 50s				
14	EPS	10s, 20s	60s, 70s	30s, 40s, 50s				
15	EPT	10s	20s, 30s, 50s, 60s, 70s	40s				
16	CD	10s, 70s	20s, 30s, 60s, 70s	20s, 50s				
17	CAP	10s, 20s	60s, 70s	30s, 40s, 50s				
18	CAO	70s	10s, 60s	20s, 30s, 50s	40s			

a, b, c, d, e, f, g: Multiple Comparison Results of Tukey Test, the same superscript on mean values indicated no significant mean differences between age groups. $\alpha = 0.05$

The phasic age-growth phenomenon of the auricle is observed in several variables. The age-group means of CAP decline significantly after 39 years of age ($p < 0.05$). The age-group means of EPS, EPT, and CAO decline significantly after 49 years of age ($p < 0.05$). By contrast, the age-group means of AIA, TH, and CD show irregular patterns without ordered age-related trends. The remaining eleven variables are classified as persistent age-growth variables (AL, AWP, AWI, CL, EML, LL, LW, CW, CCL, CCW, and TL). Among these, AL demonstrates a consistently increasing trend across all seven age-group comparisons.

The phasic age-growth phenomenon of the external acoustic meatus (EAM) follows a similar pattern. The age group means of EW, FBL, FBW, FBC, FBA, and ES-II decline significantly after 59 years of age ($p < 0.05$). The age-group means of SBW and SBA decline significantly after 49 years of age ($p < 0.05$), while the age-group means of SBC decline significantly after 69 years of age ($p < 0.05$). The remaining thirteen variables are identified as persistent age-growth variables (EL, EC, EA, SBL, SD-I, SD-II-F, SD-II-E, ED-P, ED-C, ACE, AS-I, ES-I, and AS-II).

Table 3.9 Multiple comparisons of characterization variables for the external acoustic meatus (by age groups)

		Multiple comparisons (age group)					
No.	Variable	a	b	c	d	e	f
19	EL	10s	20s	30s	40s, 50s	60s, 70s	
20	EW	10s	20s, 60s, 70s	30s, 40s, 50s			
21	EC	10s	20s	30s, 40s, 50s	60s	70s	
22	EA	10s	20s	30s, 40s,	50s, 60s, 70s		
23	FBL	10s	20s	30s, 40s, 60s, 70s	50s		
24	FBW	10s	20s	30s, 60s, 70s	40s, 50s		
25	FBC	10s	20s	30s, 40s, 60s, 70s	50s		
26	FBA	10s	20s	30s, 40s, 60s, 70s	50s		
27	SBL	10s	20s, 30s, 40s, 50s, 60s, 70s				
28	SBW	10s	20s,	30s, 50s, 60s, 70s	40s		
29	SBC	10s	20s	30s, 50s, 60s, 70s	40s, 50s, 60s		
30	SBA	10s	20s	30s, 50s, 60s, 70s	40s		
31	SD-I	10s	20s, 30s	40s, 50s	60s, 70s		
32	SD-II-F	10s	20s, 30s, 40s, 50s	30s, 40s, 50s, 60s	70s		
33	SD-II-E	10s	20s, 30s	40s, 50s	60s	70s	
34	ED-C	10s	20s, 30s, 40s	50s, 60s, 70s			
35	ED-P	10s	20s	30s, 40s	50s, 60s, 70s		
36	ACE	10s	20s	30s, 40s	50s	60s	70s
37	AS-I	10s, 20s, 30s	20s, 30s, 40s	50s, 60s, 70s			
38	ES-I	10s, 20s	30s, 40s	50s, 60s	70s		
39	AS-II	10s, 20s, 30s, 40s, 50s	30s, 40s, 50s, 60s	70s			
40	ES-II	10s, 20s	20s, 30s	60s, 70s	40s, 50s		

a, b, c, d, e, f: Multiple Comparison Results of Tukey Test, the same superscript on mean values indicated no significant mean differences between age groups. $\alpha = 0.05$

3.3.3 *Bilateral and Sexual Differences*

The results of sexual and bilateral differences in measurement variables are shown in Tables 3.10 and 3.11. Among the 18 auricular measurement variables, all except LW, CCL, TL, and TH exhibit significant sexual differences, with males having larger mean values ($p < 0.001$). Except for CAO, the left-side means are larger for the remaining variables, but no bilateral differences are significant ($p > 0.05$). For sex differences, the largest linear difference is observed in EML (mean difference between males and females = 1.79 mm), and the largest angular difference is

Table 3.10 Comparison of difference of male-female and left-right in characterization variables for the auricle

		Male-female (independent samples)						Left-right (paired samples)			
No.	Variable	Gender	Mean	SD	Gender	Mean	SD	Gender	Mean	SD	Gender
1	AL	Male	66.32	5.50	1.40*	0.89*		Left	66.23	5.60	0.22
	(mm)	Female	64.92	5.69				Right	66.01	5.59	
2	AWP	Male	35.95	3.57	0.93*	0.71*		Left	35.84	3.48	0.20
	(mm)	Female	35.02	3.42				Right	35.64	3.52	
3	AWI	Male	28.33	2.59	0.81*	0.52*		Left	28.32	2.58	0.19
	(mm)	Female	27.52	2.62				Right	28.13	2.62	
4	LL	Male	19.94	3.05	0.98*	0.21		Left	20.03	3.04	0.17
	(mm)	Female	18.96	2.99				Right	19.87	3.01	
5	LW	Male	19.11	2.49	0.03			Left	19.22	2.45	0.14
	(mm)	Female	19.08	2.41				Right	19.08	2.45	
6	EML	Male	56.83	5.34	1.79*	0.74		Left	56.59	5.55	0.32
	(mm)	Female	55.04	5.66				Right	56.27	5.48	
7	AIA (°)	Male	19.39	4.67	0.72*	0.57		Left	19.03	4.48	0.01
		Female	18.67	4.24				Right	19.02	4.46	
8	CL	Male	29.52	3.18	0.46*	0.09		Left	29.56	3.22	0.21
	(mm)	Female	29.06	3.09				Right	29.36	3.03	
9	CW	Male	19.73	2.21	0.52*	0.11		Left	19.72	2.18	0.21
	(mm)	Female	19.21	2.18				Right	19.52	2.22	
10	CCL	Male	18.28	1.85	−0.26			Left	18.51	1.86	0.20
	(mm)	Female	18.54	1.90				Right	18.31	1.89	
11	CCW	Male	15.55	2.24	0.51*	0.34		Left	15.44	2.20	0.14
	(mm)	Female	15.04	2.17				Right	15.30	2.22	
12	TL	Male	18.18	1.85	0.01			Left	18.35	1.89	0.21
	(mm)	Female	18.17	1.89				Right	18.15	1.78	
13	TH	Male	8.31	1.30	0.09			Left	8.45	1.25	0.17
	(mm)	Female	8.22	1.19				Right	8.29	1.23	
14	EPS	Male	17.06	2.35	0.87*	0.48*		Left	16.76	2.52	0.26
	(mm)	Female	16.19	2.45				Right	16.50	2.37	
15	EPT	Male	24.48	3.15	1.27*	0.60*		Left	23.95	3.30	0.23
	(mm)	Female	23.21	3.36				Right	23.73	3.14	
16	CD	Male	16.45	2.47	0.53*	−0.26		Left	16.41	2.43	0.21
	(mm)	Female	15.92	2.31				Right	16.20	2.36	
17	CAP (°)	Male	43.57	4.78	3.33*	2.06*		Left	42.08	5.23	0.33
		Female	40.24	5.43				Right	41.75	5.52	
18	CAO (°)	Male	59.28	12.40	3.84*	−0.27		Left	57.17	12.04	−0.38
		Female	55.44	11.74				Right	57.55	12.42	

*$p < 0.05$, $\alpha = 0.05$, Significant values are marked with the grey colour. a: Covariates are evaluated at the following values: Stature (mean) = 1.67, Weight (mean) = 63.90. Adjustment for multiple comparisons: Bonferroni

Table 3.11 Comparison of difference of male-female and left-right in characterization variables for the external acoustic meatus

No.	Variable	Male-female (independent samples)					Left-right (paired samples)			
		Gender	Mean	SD	Mean difference	Adjusted difference[a]	Ear side	Mean	SD	Mean difference
19	EL (mm)	Male	15.71	1.13	0.71*	0.68*	Left	15.43	1.34	0.15*
		Female	15.00	1.55			Right	15.28	1.34	
20	EW (mm)	Male	8.70	0.84	0.18*	0.13	Left	8.61	0.90	0.01
		Female	8.52	0.92			Right	8.60	0.86	
21	EC (mm)	Male	41.00	2.85	1.71*	1.45*	Left	40.26	3.18	0.35*
		Female	39.29	3.65			Right	39.91	3.32	
22	EA (mm^2)	Male	107.87	14.75	6.14*	3.25*	Left	105.37	15.50	1.12
		Female	101.73	16.51			Right	104.25	15.76	
23	FBL (mm)	Male	10.00	1.06	0.62*	0.48*	Left	9.74	1.21	0.11
		Female	9.38	1.32			Right	9.64	1.17	
24	FBW (mm)	Male	7.82	0.90	0.25*	0.02	Left	7.69	0.93	−0.01
		Female	7.57	0.98			Right	7.70	0.95	
25	FBC (mm)	Male	28.68	2.58	1.38*	0.89*	Left	28.07	2.79	0.16
		Female	27.30	2.92			Right	27.91	2.71	
26	FBA (mm^2)	Male	61.20	10.94	5.36*	3.00*	Left	58.78	11.12	0.51
		Female	55.84	11.12			Right	58.27	10.94	
27	SBL (mm)	Male	8.10	0.65	0.38*	0.28*	Left	7.95	0.76	0.08
		Female	7.72	0.89			Right	7.88	0.78	
28	SBW (mm)	Male	7.25	0.73	−0.001		Left	7.24	0.86	−0.02
		Female	7.25	1.01			Right	7.26	0.88	
29	SBC (mm)	Male	24.53	1.95	0.73*	0.21	Left	24.26	2.07	0.09
		Female	23.80	2.23			Right	24.17	2.11	
30	SBA (mm^2)	Male	46.20	6.80	1.70*	0.26	Left	45.48	7.34	0.30
		Female	44.46	7.72			Right	45.18	7.18	
31	SD-I (mm)	Male	7.65	1.25	0.29*	0.001	Left	7.54	1.38	0.07
		Female	7.36	1.27			Right	7.47	1.14	
32	SD-II-F (mm)	Male	7.18	1.22	−0.20*	−0.31*	Left	7.28	1.33	0.01
		Female	7.38	1.50			Right	7.27	1.39	
33	SD-II-E (mm)	Male	13.62	1.90	0.16		Left	13.56	2.05	0.08
		Female	13.46	2.30			Right	13.49	2.15	
34	ED-C (mm)	Male	13.92	1.39	−0.35*	−0.70*	Left	14.09	1.45	−0.04
		Female	14.27	1.57			Right	14.13	1.51	
35	ED-P (mm)	Male	10.83	1.31	−0.26*	−0.56*	Left	10.96	1.35	−0.01
		Female	11.09	1.41			Right	10.97	1.37	
36	ACE (°)	Male	24.82	2.99	−0.02		Left	24.82	3.02	−0.02
		Female	24.84	3.13			Right	24.84	3.10	

(continued)

Table 3.11 (continued)

		Male-female (independent samples)					Left-right (paired samples)			
No.	Variable	Gender	Mean	SD	Mean difference	Adjusted difference[a]	Ear side	Mean	SD	Mean difference
37	AS-I (°)	Male	109.26	6.24	1.86*	2.25*	Left	108.39	6.41	−0.03
		Female	107.40	6.46			Right	108.42	6.29	
38	ES-I (°)	Male	15.32	5.71	0.14		Left	15.33	5.39	0.14
		Female	15.18	5.13			Right	15.19	5.45	
39	AS-II (°)	Male	134.80	9.19	1.73*	2.10*	Left	134.10	8.67	0.30
		Female	133.07	8.45			Right	133.80	8.97	
40	ES-II (°)	Male	30.32	6.54	0.32		Left	30.32	6.61	0.36
		Female	30.00	6.48			Right	29.97	6.41	

*$p < 0.05$, $\alpha = 0.05$, Significant values are marked with the grey colour. a: Covariates are evaluated at the following values: Stature (mean) = 1.67, Weight (mean) = 63.90. Adjustment for multiple comparisons: Bonferroni

observed in CAO (mean difference = 3.84°). For bilateral differences, the largest linear difference is EML (mean difference = 0.33 mm), and the largest angular difference is CAO (mean difference = −0.38°).

Among the 22 external acoustic meatus measurement variables, all except SBW, SD-II-E, ACE, ES-I, and ES-II show significant sexual differences ($p < 0.001$). Except for EL and EC, the remaining variables exhibit no significant bilateral differences, although their left-side means are generally larger ($p > 0.05$). For the five variables without significant sexual differences, the male values are still greater than the female means. Among the 17 variables with significant sex differences, all but SD-II-F, ED-C, and ED-P show larger mean values in males ($p < 0.001$). For sex differences, the largest linear difference is EL (mean difference between males and females = 0.71 mm), the largest angular difference is AS-I (mean difference = 1.86°), the largest circumferential difference is EC (mean difference = 1.71 mm), and the largest areal difference is EA (mean difference = 6.14 mm^2). For bilateral differences, the largest linear difference is EL (mean difference = 0.15 mm), the largest angular difference is ES-II (mean difference = 0.36°), the largest circumferential difference is EC (mean difference = 0.35 mm), and the largest areal difference is EA (mean difference = 1.12 mm^2).

After adjusting the difference values of all variables with significant sexual differences to exclude the effects of height and weight, 18 variables still show significant sex differences: AL, AWP, AWI, EPS, EPT, CAP, EL, EC, EA, FBL, FBC, FBA, SBL, SD-II-F, ED-C, ED-P, AS-I, and AS-II. In the subsequent analysis, the overall means of these variables will not be used; instead, either male or female mean values will be selected as needed to construct specific models.

3.4 Factors Associated with Shape Variations and Growth Trajectories

In this section, variables were grouped on the basis of the results of factor analysis according to between-subject variation trends, and Factors Associated with Shape Variations (FASVs) were subsequently constructed. The weights of the FASVs, as well as the weights of the measurement variables within each factor, were then calculated, through which an extended expression of Eqs. (2.1) and (2.3) was obtained. Furthermore, growth trajectories of the auricle were described from the perspective of cross-sectional data across age groups by means of the constructed FASVs.

3.4.1 Factors Associated with Shape Variations

A total of six auricular factors of age-structure variability (FASVs, eigenvalue >1) are extracted from the 18 auricular measurement variables:

1. Auricular Contour Factor (FASV 1). Seven measurement variables load on this factor, representing variations in the outer contour dimensions: AL, AWP, AWI, EML, LL, CL, and LW.
2. Tragus-Concha Length Factor (FASV 2). Two measurement variables, CCL and TL, represent variations in the contour length of the tragus-concha region.
3. Conchal Width Factor (FASV 3). Two measurement variables, CW and CCW, capture variations in the width of the concha.
4. Auricular Upper Protrusion Factor (FASV 4). Two measurement variables, EPS and CAP, reflect variations in the upper protrusion of the auricle.
5. Auricular Posterior Protrusion Factor (FASV 5). Two measurement variables, EPT and CAO, describe variations in the posterior protrusion of the auricle.
6. Non-age-related Factor (FASV 6). Three measurement variables, TH, AIA, and CD, are grouped together, as they show no significant correlation with age (Table 3.12).

A total of seven external acoustic meatus FASVs (eigenvalue >1) are constructed from the 22 external acoustic meatus measurement variables, namely:

7. EAM First Bend Factor (FASV 7). Four measurement variables, FBL, FBW, FBC, and FBA, are included in this factor, describing variations in the morphology and dimensions of the first bend of the external acoustic meatus.
8. EAM Second Bend Factor (FASV 8). Four measurement variables, SBL, SBW, SBC, and SBA, are included, reflecting variations in the morphology and dimensions of the second bend of the meatus.
9. EAM Entrance Factor (FASV 9). Four measurement variables, EL, EW, EC, and EA, are grouped to describe variations in the morphology and dimensions of the EAM entrance.

Table 3.12 FASVs of the auricle

No.	Variable	Factor loadings[a] (items with loadings >0.40)					
		FASV 1	FASV 2	FASV 3	FASV 4	FASV 5	FASV 6
		Auricular contour factor	Tragus-concha length factor	Conchal width factor	Auricular upper protrusion factor	Auricular posterior protrusion factor	Non-age-related factor
1	AL	0.90					
6	EML	0.84					
4	LL	0.77					
2	AWP	0.77					
3	AWI	0.71					
8	CL	0.60					
5	LW	0.44					
10	CCL		0.97				
12	TL		0.96				
11	CCW			0.95			
9	CW			0.92			
14	EPS				0.93		
17	CAP				0.92		
18	CAO					0.91	
15	EPT					0.86	
13	TH						0.68
7	AIA						0.60
16	CD						0.55
Eigenvalue		4.95	2.45	1.74	1.48	1.31	1.09
% of variance		27.48	13.60	9.64	8.24	7.26	6.03
Cumulative %		27.48	41.07	50.71	58.96	66.21	72.25
Items		7	2	2	2	2	3

Overall % of Variance = 72.25
KMO = 0.71, Bartlett's test sig. <0.001
[a] Rotation converged in 5 iterations

10. EAM Depth Factor (FASV 10). Three measurement variables, SD-I, SD-II-F, and SD-II-E, are included, capturing variations in the triangular morphology and depth of EAM sections.
11. EAM Entrance Position Factor (FASV 11). Three measurement variables, ACE, ED-P, and ED-C, describe variations in the positional characteristics of the EAM entrance.
12. EAM Elevation Angle Factor (FASV 12). Two measurement variables, ES-I and ES-II, reflect variations in the angular relationship between the isthmus of the meatus and the auricular horizontal plane.
13. EAM Azimuth Angle Factor (FASV 13). Two measurement variables, AS-I and AS-II, characterize variations in the azimuth angles of the EAM canal sections (Table 3.13).

Table 3.13 FASVs of the external acoustic meatus

No.	Variable	Factor loadings[a] (items with loadings >0.40)						
		FASV 7	FASV 8	FASV 9:	FASV 10	FASV 11	FASV 12	FASV 13
		EAM first bend factor	EAM second bend factor	EAM entrance factor	EAM depth factor	EAM entrance position factor	EAM elevation angle factor	EAM azimuth angle factor
26	FBA	0.94						
25	FBC	0.87						
24	FBW	0.80						
23	FBL	0.69						
29	SBC		0.94					
30	SBA		0.93					
27	SBL		0.78					
28	SBW		0.62					
22	EA			0.92				
21	EC			0.83				
20	EW			0.76				
19	EL			0.63				
33	SD-II-E				0.91			
32	SD-II-F				0.85			
31	SD-I				0.65			
34	ED-C					0.86		
35	ED-P					0.85		
36	ACE					0.49		
40	ES-II						0.71	
38	ES-I						0.55	
39	AS-II							0.82
37	AS-I							0.50
Eigenvalue		7.35	2.23	2.06	1.55	1.35	1.17	1.09
% of variance		31.96	9.69	8.95	6.72	5.88	5.08	4.72
Cumulative %		31.96	41.65	50.60	57.32	63.20	68.28	72.99
Items		4	4	4	3	3	2	2

Overall % of variance = 72.99
KMO = 0.73, Bartlett's test sig. <0.001
[a] Rotation converged in 6 iterations

According to Eq. (2.12) and the results of the variance contribution rates, the factors with higher overall variance contributions exert a greater influence on the total variability. Therefore, the two terms P_{Aur} and P_{EAM} in Eq. (2.13) can be further expressed using Eq. (2.4).

$$P_{Aur} = 0.38p_{FASV1} + 0.19p_{FASV2} + 0.13p_{FASV3} + 0.11p_{FASV4} + 0.1p_{FASV5} + 0.08p_{FASV6} \quad (3.4)$$

$$P_{EAM} = 0.44p_{FASV7} + 0.13p_{FASV8} + 0.12p_{FASV9} + 0.09p_{FASV10} + 0.08p_{FASV11} + 0.07p_{FASV12} + 0.07p_{FASV13} \quad (3.5)$$

where, P_{Aur} is the user's comfort perception associated with auricular structural factors; P_{EAM} is the user's comfort perception associated with external acoustic meatus structural factors; p is the user's comfort perception regarding variations in the Factors Associated with Shape Variations (FASVs).

Based on the construction results of the Factors Associated with Shape Variations (FASVs), the basic structure of the user's comfort perception model regarding variations in the FASVs is given in Eq. (3.5). In combination with Eq. (2.4), the complete structure of Eq. (2.1) is presented in Eq. (3.6).

$$p_{ik} = \sum_{h=1}^{l} u_{ik,h} g_{ik,h} \quad (3.6)$$

$$P_{Fitting} = \sum_{i=1}^{m} W_i \left(\sum_{k=1}^{n} \omega_{ik} \left(\sum_{h=1}^{l} u_{ik,h} g_{ik,h} \right) \right) \quad (3.7)$$

where, l is the number of measurement variables contained in the k-th subfactor of the i-th ergonomic factor; n is the number of quantitative subfactors within the i-th ergonomic factor; m is the total number of ergonomic factors; h is the h-th measurement variable within the k-th subfactor; p denotes the user's comfort perception regarding changes in the quantitative subfactor; g denotes the user's comfort value regarding changes in the measurement variables within the quantitative subfactors; W is the weight of the ergonomic factor; ω is the weight of the quantitative subfactor; u is the weight of the measurement variable.

The weights of measurement variables within a single Factor Associated with Shape Variations (FASV) are calculated using the factor evaluation coefficient weights. Specifically, factor analysis is performed on the variables contained in each FASV; the comprehensive evaluation coefficient for each measurement variable is calculated using the evaluation coefficient formula (Eq. (3.7)), and the weight of each measurement variable is then determined using the evaluation coefficient weight formula (Eq. (3.8)). Since the factor analysis of the measurement variables within a single FASV yields only one component, and the variance contribution rate equals the cumulative contribution rate, Eq. (3.7) can be further simplified to Eq. (3.9). The results of the variable weights are presented in Table 3.14.

Table 3.14 Weight coefficients of FASVs

Auricular FASVs	No.	Variable	Weight coefficient ω	EAM FASVs	No.	Variable	Weight coefficient ω
FASV 1	1	AL	0.17	FASV 7	26	FBA	0.23
	6	EML	0.13		25	FBC	0.20
	4	LL	0.14		24	FBW	0.28
	2	AWP	0.15		23	FBL	0.29
	3	AWI	0.10	FASV 8	29	SBC	0.25
	8	CL	0.17		30	SBA	0.20
	5	LW	0.13		27	SBL	0.27
FASV 2	10	CCL	0.50		28	SBW	0.28
	12	TL	0.50	FASV 9	22	EA	0.21
FASV 3	11	CCW	0.50		21	EC	0.22
	9	CW	0.50		20	EW	0.28
FASV 4	14	EPS	0.50		19	EL	0.29
	17	CAP	0.50	FASV 10	33	SD-II-E	0.32
FASV 5	18	CAO	0.50		32	SD-II-F	0.30
	15	EPT	0.50		31	SD-I	0.38
FASV 6	13	TH	0.24	FASV 11	34	ED-C	0.32
	7	AIA	0.37		35	ED-P	0.30
	16	CD	0.39		36	ACE	0.38
				FASV 12	40	ES-II	0.50
					38	ES-I	0.50
				FASV 13	39	AS-II	0.50
					37	AS-I	0.50

$$Coefficient_{ik,h} = \frac{\sum_{k=1}^{n}\left(\frac{FacorLoading_{ik,h}}{\sqrt{Eigen_{ik}}} \times (\%of\ Variance)_{ik}\right)}{\sum_{k=1}^{n}(\%of\ Variance)_{ik}} \tag{3.8}$$

$$u_{ik,h} = \frac{Coefficient_{ik,h}}{\sum_{h=1}^{l} Coefficient_{ik,h}} \tag{3.9}$$

$$Coefficient_{ik,h} = \frac{FacorLoading_{ik,h}}{\sqrt{Eigen_{ik}}} \tag{3.10}$$

where, l is the number of measurement variables contained in the k-th subfactor of the i-th ergonomic factor; n is the number of quantitative subfactors within the i-th ergonomic factor; m is the total number of ergonomic factors; h is the h-th measurement variable within the k-th subfactor; u is the weight of the measurement variable.

3.4.2 Growth Trajectories

Figures 3.10 and 3.11 illustrate the visual representations of FASV 1 to FASV 13. Based on the findings in Sect. 3.3 and the construction results of the Factors Associated with Shape Variations (FASVs), the growth trajectories of the auricle from a cross-sectional perspective of age groups are summarized.

The morphological dimensions of the auricular outer contour demonstrate clear age-related changes, with both length and width increasing across the lifespan. Notably, auricular length is the only variable that exhibits significant growth across every pairwise comparison between adjacent age groups. The lengths of the concha and tragus also increase significantly between ages 20–29 and 40–49, but both show marked declines between ages 70–79. Similarly, the widths of the concha and the cavum concha follow this trend, increasing significantly during early and middle adulthood and declining significantly after age 50, with particularly pronounced reductions observed between ages 50–59 and 70–79.

Patterns of auricular protrusion reveal further nuances. The level of upper protrusion grows significantly between ages 30–39 but subsequently declines after age 60, returning by ages 70–79 to levels comparable with those observed during late adolescence (15–19 years). Posterior protrusion demonstrates a different trajectory: it declines progressively across the age intervals 50–59, 60–69, and 70–79, reaching

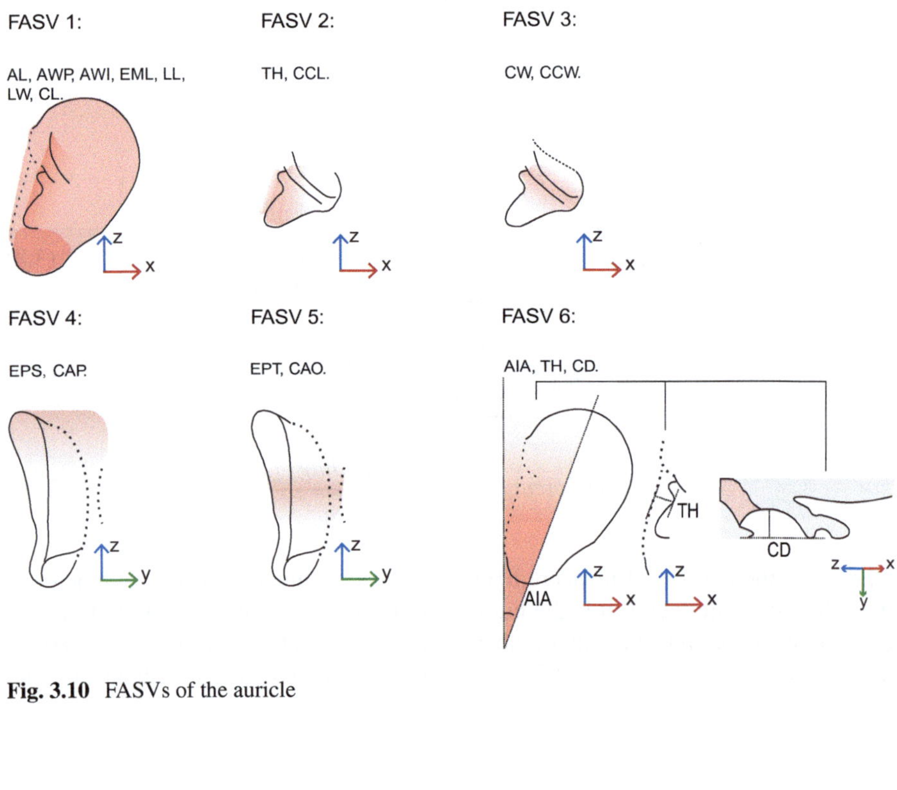

Fig. 3.10 FASVs of the auricle

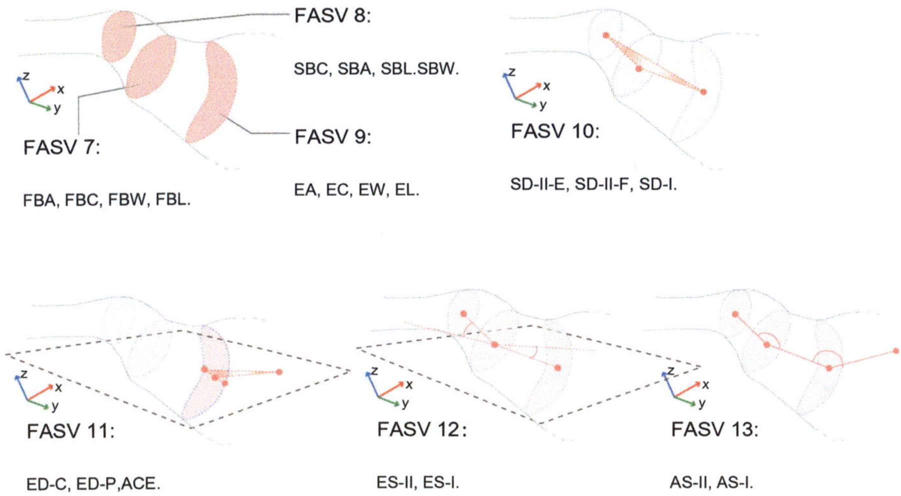

Fig. 3.11 FASVs of the external acoustic meatus

its lowest angular values in the oldest age group and approximating the level of 15–19 years. By contrast, the auricular inclination angle, tragus height, and conchal depth show no systematic trends in variation from ages 15 to 79, suggesting greater structural stability in these dimensions.

Within the external acoustic meatus (EAM), the first bending cross-section decreases significantly in later adulthood, particularly between ages 50–79. The second bend length demonstrates significant variation in every decade after age 29, whereas its width, circumference, and area decline consistently after age 39. The entrance cross-section of the EAM shows a different dynamic: its length, circumference, and area increase continuously with age, while its width declines significantly between ages 60–79.

The depth of the EAM increases progressively with age, reflecting a gradual smoothing trend and the potential for the first bend to be completely smoothed in older age groups. Correspondingly, the mean values of the meatus entrance position variables increase with age. In line with the findings reported in Sect. 3.3, the growth of the tragus and concha indicates that the position of the EAM entrance gradually shifts toward the craniofacial coronal plane as age advances (Fig. 3.12).

Angular measures further highlight the trajectory of structural transformation. The elevation angle of the EAM increases steadily with age, suggesting a progressive upward shift of its relative position within the cranial structure toward the parietal and occipital bones (Fig. 3.12). At the same time, the orientation angle of the isthmus increases continuously, indicating that both the first and second bends of the EAM follow a gradual smoothing process with the potential for eventual complete straightening.

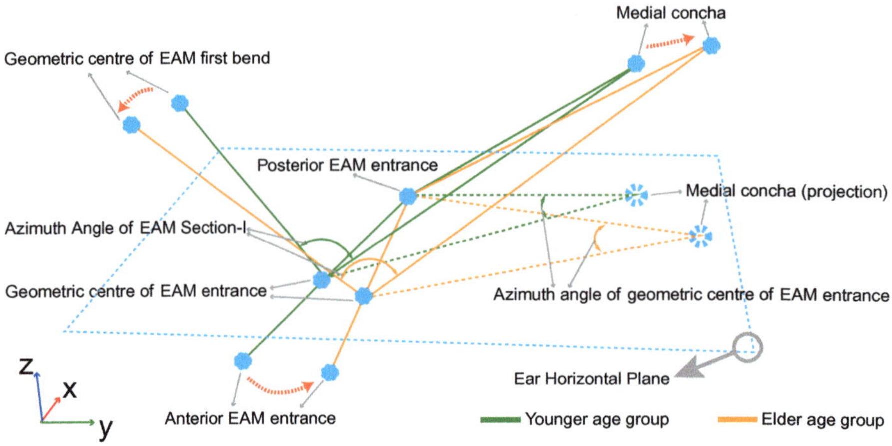

Fig. 3.12 Variation of entrance position of the external acoustic meatus from younger to elder age groups

3.5 Structural and Ergonomic Implications of External Ear Characterization

A comprehensive methodological framework is developed for the characterization of the external ear, comprising both the auricle and the external acoustic meatus (EAM). Forty characterization variables are defined, including eighteen auricular variables and twenty-two EAM variables, of which twenty-three are newly introduced. The framework integrates anatomical landmarking, orientation and reference plane construction, variable definition, three-dimensional data acquisition and processing, and statistical modelling. Beyond descriptive measurement, the analysis highlights age-related trajectories, sex differences, and design thresholds, with implications for ergonomic design and predictive modelling.

3.5.1 Age Associations and Functional Implications

Of the forty variables, thirty-eight demonstrate significant associations with age, while only AIA, TH, and CD show no clear relationship. This indicates that the external ear is highly dynamic across the lifespan, with both the auricle and the EAM undergoing measurable morphological transformation.

For the auricle, outer contour dimensions expand continuously, as both length and width increase with age. Particularly noteworthy is the growth of lobular length, which rises by approximately 32%. This expansion is attributed less to cultural practices and more to the biological decline of tissue elasticity with age. Conchal and tragal lengths exhibit phasic age-growth: they increase significantly during the twenties and forties, but decline during the seventies. Similarly, cavum concha width follows a comparable pattern, with additional midlife increases observed in the fifties. Protrusion measures also follow stage-dependent trends: upper auricular protrusion rises in the thirties but falls after the sixties, while posterior protrusion declines steadily after the fifties, returning in older groups to levels observed in late adolescence.

Within the EAM, age-related change manifests differently. Entrance variables, including length, circumference, and area, grow continuously with age, while width declines in late adulthood. In contrast, the cross-sectional areas of the first and second bends diminish after midlife, indicating narrowing in deeper canal regions. Depth measures, such as SD-I, SD-II-F, and SD-II-E, increase progressively with age, reflecting gradual straightening of the canal. Orientation variables also increase, suggesting that the EAM shifts cranially and medially with advancing age. Collectively, these findings reveal a dual pattern: while the auricle tends to enlarge externally, the canal undergoes internal reconfiguration, marked by straightening and reorientation.

3.5.2 Sexual Differences and Design Principles

Eighteen variables retain significant sex differences even after controlling for height and weight. These differences cluster around the tragus, concha, and lobule, demonstrating that sex affects both size and localized morphology. For ergonomic purposes, it recommends that when sex differences are present, female values should serve as design references. This ensures inclusivity and prevents the risk of discomfort that may result from adopting larger male-based thresholds.

Bilateral asymmetry is minimal in this dataset, unlike findings in some international studies. This suggests that left-right differentiation may not be necessary in general design for the Chinese case study. However, the contrast underscores the importance of population-specific validation before applying ergonomic data universally.

3.5.3 Growth Models and Lifespan-Oriented Design

While earlier research often assumed a simple linear pattern, the findings show that auricular and EAM variables follow diverse trends. Some variables, such as auricular length, auricular width, and lobular length, exhibit persistent and monotonic growth throughout the lifespan, showing no evidence of plateau or decline. Others, particularly conchal and tragal dimensions and several EAM measures, follow phasic or nonlinear trajectories characterized by periods of increase, stabilization, and eventual reduction in older age. These differentiated models more accurately reflect the underlying biological mechanisms of ear aging. Persistent growth is associated with gradual loss of tissue elasticity and continued expansion of soft tissues, while nonlinear patterns reflect cartilage remodelling, collagen changes, and localized structural degeneration.

From an ergonomic perspective, recognizing both persistent and nonlinear growth trajectories enables more precise predictive frameworks for lifespan-oriented design. Products such as industrial ear protectors, medical hearing devices, and assistive wearables often remain in use for many years. Designing these products with reference to persistent growth variables ensures long-term fit, while accounting for nonlinear variables allows developers to anticipate specific points of fit degradation. This dual approach shifts ergonomics from static design toward anticipatory design, where age-related change, whether continuous or stage-specific, is treated as a central design parameter.

3.5.4 Applications and Broader Relevance

The forty defined variables fill critical gaps in existing ISO and national standards, which often omit three-dimensional ear data. These measures can serve as benchmarks for future revisions, ensuring that standards align with contemporary anthropometric diversity.

Accurate auricular and EAM data enhance research in acoustics, by improving models of sound propagation; in medicine, by supporting surgical navigation and prosthesis design; and in immersive technologies, by optimizing comfort and coupling in head-mounted displays.

With advances in 3D scanning and AI-based modelling, the defined variables provide a structured dataset for individualized design. Algorithms trained on this framework can generate customized fittings or even support on-demand 3D printing of ear-related devices.

3.5.5 Evidence Scope, Measurement Constraints, and Applicability

The representativeness of the present dataset is conditioned by the demographic composition of the domestic population. China is multi-ethnic, yet Han Chinese constitute over 90% of the total population. In this program, participants from 12 ethnic groups were included, covering approximately 83% Han. This structure permits robust characterization of population-level morphology, but subgroup-specific features remain underrepresented, limiting direct applicability to fine-grained cross-ethnic comparative design. Future extensions should expand subgroup coverage to address this limitation.

The measurement-variable framework reflects current literature gaps and human-product interaction needs. While comprehensive, it is not exhaustive. Emerging technologies or application contexts may introduce requirements for new descriptors, such as metrics for a third canal segment or scaphoid fossa, which would extend beyond the present set. Moreover, factor analysis yielded cumulative variance contributions below 80%, suggesting that explanatory sufficiency remains constrained by subgroup coverage and extended recruitment windows (>12 months). Denser sampling with shorter cycles may provide stronger explanatory power. Health-related exclusions also shape the evidence base. Over 70% of respondents under 20 years reported frequent or long-term analgesic use, while over 60% of respondents above 60 years reported head-facial pain. Both conditions could alter discomfort thresholds or induce hyperalgesia, and thus these groups were excluded from exposure-response modelling. For interpretive continuity, results from neighbouring age groups were applied to estimate their potential distributions. This assumption provides continuity for modelling but should be treated as provisional until new recruitment conditions allow direct sampling. Another methodological constraint arises from measurement safety. The comfort-related deformation ratios of tragus, cavum, and EAM were derived indirectly rather than measured continuously, owing to risks associated with invasive procedures. Future developments in safe in vivo measurement technologies will be required for direct empirical validation.

Competing Interests This study is by "the Fundamental Research Funds for the Central Universities" (NPU Grant No. D5000250035, SEU Grant No. 2242024S30013). Additional support comes from the Northwestern Polytechnical University. The authors have no conflicts of interest to declare that are relevant to the content of this chapter.

References

1. Fan, H., Wang, M., Xu, S., He, F., Zhang, X., Shi, J., & Chai, C. (2025). Surface pressure sensitivity of the auricle and periauricular areas for design purposes of product interaction. *Proceedings of the Human Factors and Ergonomics Society Annual Meeting, 69*(1). https://doi.org/10.1177/10711813251367739
2. Fan, H., Wang, M., Xu, S., Shi, J., Jiang, A., Ga, F., & Chai, C. (2025). Chinese auricular shape types for ear-related wearables: A study of 1,400 participants from 2016 to 2020. *Proceedings of the Human Factors and Ergonomics Society Annual Meeting, 58*, 58–61. https://doi.org/10.1177/10711813251367747
3. Alexander, M., & Laubach, L. L. (1968). *Anthropometry of the human ear (a photogrammetric study of USAF flight personnel). AMRL-TR-68-1*. Aerospace Medical Research Laboratory.
4. Jung, H. S., & Jung, H. S. (2003). Surveying the dimensions and characteristics of Korean ears for the ergonomic design of ear-related products. *International Journal of Industrial Ergonomics, 31*(6), 361–373. https://doi.org/10.1016/S0169-8141(02)00237-8
5. Stavrakos, S. K., Ahmed-Kristensen, S., & Goldman, T. (2016). Using archetypes to create user panels for usability studies: Streamlining focus groups and user studies. *Applied Ergonomics, 56*, 108–116. https://doi.org/10.1016/j.apergo.2016.02.016

6. Stavrakos, S. K., & Ahmed-Kristensen, S. (2016). Methods of 3D data applications to inform design decisions for physical comfort. *Work, 55*(2), 321–334. https://doi.org/10.3233/WOR-162399
7. Lee, W., Jung, H., Bok, I., et al. (2016). Measurement and application of 3D ear images for earphone design. *Proceedings of the Human Factors and Ergonomics Society Annual Meeting, 60*(1), 1053–1057. https://doi.org/10.1177/1541931213601243
8. Roebuck, J. A., & Casali, J. G. (2011). Re-inventing anthropometry for design of ear-mounted or ear-coupled products. In *Proceedings of the human factors and ergonomics society* (pp. 1649–1653). https://doi.org/10.1177/1071181311551344
9. Roebuck, J. A. (2015). Fixing flaws in engineering anthropometry, ears and all. In *Proceedings of the human factors and ergonomics society annual meeting* (Vol. 59, pp. 1414–1418). SAGE. https://doi.org/10.1177/1541931215591307
10. Roebuck, J. A. (2013). Advances in anthropometric accounting for ear digital modelling. In *Proceedings of the 4th international conference on 3D body scanning technologies* (pp. 360–371). https://doi.org/10.15221/13.360
11. Fan, H., Yu, S., Chu, J., et al. (2019). Anthropometric characteristics and product categorization of Chinese auricles for ergonomic design. *International Journal of Industrial Ergonomics, 69*(1), 118–141. https://doi.org/10.1016/j.ergon.2019.01.004
12. Standring, S. (2016). *Gray's anatomy: The anatomical basis of clinical practice* (41st ed.). Elsevier.
13. Purkait, R., & Singh, P. (2007). Anthropometry of the normal human auricle: A study of adult Indian men. *Aesthetic Plastic Surgery, 31*(4), 372–379. https://doi.org/10.1007/s00266-006-0231-4
14. Liu, B. S. (2008). Incorporating anthropometry into design of ear-related products. *Applied Ergonomics, 39*(1), 115–121. https://doi.org/10.1016/j.apergo.2006.12.005
15. Wang, B., Dong, Y., Zhao, Y., et al. (2011). Computed tomography measurement of the auricle in Han population of north China. *Journal of Plastic, Reconstructive & Aesthetic Surgery, 64*(1), 34–40. https://doi.org/10.1016/j.bjps.2010.03.053
16. Barnes, J., Sabo, R. T., & Coelho, D. H. (2018). A novel method to measure the external auditory canal: Normative data and practical implications. *American Journal of Otolaryngology - Head and Neck Medicine and Surgery, 39*(2), 146–149. https://doi.org/10.1016/j.amjoto.2017.12.006
17. Farkas, L. G., & Posnick, J. C. (1992). Growth and development of regional units in the head and face based on anthropometric measurements. *Cleft Palate-Craniofacial Journal, 29*(4), 301–302. https://doi.org/10.1597/1545-1569_1992_029_0301_gadrua_2.3.co_2
18. Lee, W., Yang, X., Jung, H., et al. (2018). Anthropometric analysis of 3D ear scans of Koreans and Caucasians for ear product design. *Ergonomics, 61*(11), 1480–1495. https://doi.org/10.1080/00140139.2018.1493150
19. Alexander, K. S., Stott, D. J., Sivakumar, B., et al. (2011). A morphometric study of the human ear. *Journal of Plastic, Reconstructive & Aesthetic Surgery, 64*(1), 41–47. https://doi.org/10.1016/j.bjps.2010.04.005
20. Ferrario, V. F., Sforza, C., Ciusa, V., et al. (1999). Morphometry of the normal human ear: A cross-sectional study from adolescence to mid-adulthood. *Journal of Craniofacial Genetics and Developmental Biology, 19*(4), 226–233.
21. Purkait, R. (2013). Progression of growth in the external ear from birth to maturity: A 2-year follow-up study in India. *Aesthetic Plastic Surgery, 37*(3), 605–616. https://doi.org/10.1007/s00266-013-0097-1
22. Zhao, S., Li, D., Liu, Z., et al. (2018). Anthropometric growth study of the ear in a Chinese population. *Journal of Plastic, Reconstructive & Aesthetic Surgery, 71*(4), 518–523. https://doi.org/10.1016/j.bjps.2017.10.010
23. Wang, M., Fan, H., Yu, S., et al. (2022). Analysis of the auricles and auricular shape types for ear-related wearables: A study of mainland Chinese samples aged 15-79. *Work, 73*(1), 1–20.
24. Yu, J. F., Lee, K. C., Wang, R. H., et al. (2015). Anthropometry of external auditory canal by non-contactable measurement. *Applied Ergonomics, 50*, 50–55. https://doi.org/10.1016/j.apergo.2015.01.008

25. Fan, H., Yu, S., Wang, M., et al. (2021). Analysis of the external acoustic meatus for ergonomic design: Part I - Measurement of the external acoustic meatus using casting, scanning and rapid estimation approaches. *Ergonomics, 64*(5), 640–656. https://doi.org/10.1080/00140139.2020.1869215
26. Fan, H., Yu, S., Wang, M., et al. (2021). Analysis of the external acoustic meatus for ergonomic design: Part II - Anthropometric variations of the external acoustic meatus by sex, age and side in Chinese population. *Ergonomics, 64*(5), 657–670. https://doi.org/10.1080/00140139.2020.1869216
27. Doutres, O., Sgard, F. C., Benacchio, S., et al. (2018). Earplug comfort: From subjective assessment on the field to objective measurement and simulation using augmented artificial heads. *The Journal of the Acoustical Society of America, 143*(3), 1910–1930. https://doi.org/10.1121/1.5025311
28. Chiou, W. K., Huang, D. H., & Chen, B. H. (2016). Anthropometric measurements of the external auditory canal for hearing protection earplug. *Adv. Intell. Syst. Comput., 491*, 163–171. https://doi.org/10.1007/978-3-319-41929-9_16
29. European Committee for Standardization. (2006). BS EN 960:2006: Headforms for use in the testing of protective helmets [Standard].
30. Fan, H., Wang, M., Adamson, C. H., et al. (2020). Anthropometric analysis of external acoustic meatus for ergonomic design. *Proceedings of the Human Factors and Ergonomics Society Annual Meeting, 64*(1), 1833–1835. https://doi.org/10.1177/1071181320641397
31. Lacko, D., Vleugels, J., Fransen, E., et al. (2017). Ergonomic design of an EEG headset using 3D anthropometry. *Aesthetic Plastic Surgery, 31*(4), 372–379. https://doi.org/10.1016/j.apergo.2016.06.002

Chapter 4
Mechanism Models of the External Ear Shape-Structure Factors

Abstract A comprehensive ergonomic and anthropometric investigation is presented into the morphological characteristics, deformation behavior, and wear comfort mechanisms of the external ear, encompassing both the auricle and the external acoustic meatus (EAM). To address the limitations of conventional static measurements, deformation ratio is introduced as a normalized metric for quantifying tissue deformation under compressive and expansive loads induced by ear-worn and in-ear devices. Through controlled deformation experiments combined with comfort perception scaling, deformation-comfort models are established to identify comfort thresholds and perceptual limits for key auricular and EAM parameters. Analysis of data from a large, demographically stratified adult population indicates that deformation ratios of major auricular dimensions are not significantly influenced by sex, age group, or laterality, supporting the use of pooled predictive models. For the EAM, where direct measurement of cross-sectional circumference and area is restricted by anatomical accessibility, rapid estimation methods based on elliptical approximation and regression fitting are proposed and validated with high accuracy. In addition, persistent and phasic age-growth models are developed to describe lifespan-related changes in auricular and canal dimensions. By integrating morphology, deformation mechanics, comfort perception, and age-related growth patterns, this chapter provides a quantitative foundation for the ergonomic design, evaluation, and long-term adaptation of ear-related wearable devices.

Keywords External ear · Deformation-comfort model · Wear comfort threshold · External acoustic meatus · Age-growth modelling

The external ear, comprising the auricle and the external acoustic meatus (EAM), represents a unique morphological structure at the interface of human biology, sensory perception, and engineered design. Unlike many other craniofacial elements, the auricle is simultaneously a biological identifier, a visual marker of individuality, and the first structural element modulating auditory input. Its morphology, while often perceived as static, is in fact dynamic, reflecting the interplay of genetic, cultural, environmental, and age-related factors. In engineering anthropometry, this duality, in which the auricle serves both as an aesthetic form and as a functional

H. Fan et al., *Engineering Anthropometry of the External Ear for Human-Product Interaction*, https://doi.org/10.1007/978-981-92-2160-8_4

gateway to the auditory system, makes it an especially critical site for ergonomic characterization and design translation.

Traditional anthropometry has largely focused on gross body dimensions, skeletal landmarks, and surface measurements that support apparel, workstation, or vehicle design. However, the rapid proliferation of wearable technologies, particularly those located on or within the ear, demands a deeper, fine-grained understanding of local anatomy. Products such as hearing aids, earphones, earplugs, bone-conduction headsets, and emerging health-monitoring devices rely on intimate contact with auricular and canal structures. Comfort, fit, acoustic sealing, and biomechanical safety all depend on accurate knowledge of these complex geometries. Engineering anthropometry must therefore evolve beyond traditional linear metrics to encompass multidimensional descriptors of ear morphology, cross-sectional geometry, and deformation behaviour.

A persistent challenge in this domain lies in measurement accessibility. The auricle, being external, allows for relatively direct observation and surface scanning. The EAM, by contrast, is embedded within the skull, follows a curved trajectory with multiple bends, and is bounded by compliant soft tissue. These features render direct measurement of cross-sectional circumference and area impractical under routine conditions. Historically, studies have relied on impression molding or radiographic imaging, both of which present limitations such as distortion caused by material shrinkage, user discomfort, and radiation exposure. Consequently, predictive estimation models have emerged as a necessary approach. By leveraging directly measurable length and width parameters as proxies for elliptical cross-sections, circumference and area can be approximated with high fidelity. Such approaches not only enable large-scale data collection but also reduce methodological variance, making results comparable across laboratories and populations.

Age represents another critical determinant in external-ear morphology. Unlike many craniofacial structures that stabilize in adulthood, the auricle continues to change throughout life. Persistent age-growth manifests as gradual, monotonic increases in dimensions such as auricular length, width, and lobule size. Phasic age-growth reflects transitional or stage-dependent changes, including non-linear trajectories of protrusion or canal cross-sectional dimensions. These dual phenomena underscore the importance of incorporating both persistent and phasic models when predicting long-term fit and comfort. From an engineering perspective, age-linked anthropometry is indispensable for universal design, where products must remain functional and comfortable across extended service lives and multiple life stages.

Comfort modelling represents a further extension of anthropometric analysis into human-product interaction. While linear dimensions provide the structural basis, it is deformation, referring to how the auricle and canal respond under load, that determines subjective comfort and safety. By formalizing deformation ratios and linking them to comfort perception values, thresholds can be established for design evaluation. This approach mirrors exposure-response functions in physiology, where quantitative thresholds translate biological responses into actionable design limits. In practice, this means that designers can calibrate device dimensions not only against static morphology but also against tolerance ranges for compression or expansion. Such integration of comfort perception into anthropometric modelling represents a significant advance in ergonomic methodology.

4.1 Deformation-Comfort Model, Wear Comfort Threshold, and Comfort Perception Limit

4.1.1 Deformation Ratio

The quantification of the External Ear deformation represents a key prerequisite for developing an ergonomic basis of ear-related wearable design. The auricle, as a morphological-structural entity, exhibits complex variations in geometry, tissue elasticity, and curvature across individuals. When external pressure or sustained mechanical load is introduced, such as from in-ear or on-ear devices, these variations give rise to measurable deformations, which, in turn, directly affect the perception of comfort. From a theoretical standpoint, auricular deformation can be regarded as a biomechanical transformation that mediates between external mechanical stimuli and subjective sensory evaluation.

To systematically capture such transformation, deformation cannot be assessed merely in terms of absolute displacement or raw dimensional change. These descriptors are highly dependent on initial anthropometric scale and do not enable effective comparison across individuals. A normalized indicator, deformation ratio (Eq. (4.1)), therefore, becomes essential. By expressing the ratio of the deformed state relative to the natural state, deformation ratio provides a dimensionless index that reveals structural responsiveness independent of individual ear size or shape. Within this framework, two conceptual objectives emerge. First, the measurement and standardization of shape-structure variables under deformation establish a reference system for quantifying the extent of structural adaptation. Second, the formulation of a quantitative relationship between deformation ratio and comfort value (degree of wear comfort) enables the identification of critical thresholds.

Such a deformation-comfort model provides a dual bridge: on one hand, it links physical deformation metrics with psychophysical perception; on the other, it offers a theoretical foundation for ergonomic design principles. By situating auricular mechanics within a normalized quantitative framework, the model contributes to the establishment of comfort-based exposure limits and design criteria for wearable products.

$$R = \frac{S_{def} - S_{ini}}{S_{ini}} \times 100\%,\ 0 < S_{ini} \le S_{def} \tag{4.1}$$

where S_{def} is the measured value of the linear variable after deformation; S_{ini} is the measured value of the linear variable in its natural state; R is the deformation ratio.

4.1.2 Methods of Deformation Characterization

Types of Deformation and Applicable Variables

When wearable devices make contact with the auricle, local deformations arise within the cartilage structure and cutaneous surface. These deformations can be broadly categorized into two primary types: compressive deformation of the auricle, typically associated with cap-type and cup-type devices (Figs. 4.1 and 4.2a, b), and expansive deformation of the tragus, concha, and external acoustic meatus, generally induced by ear-insert elements of band-type or plug-type devices (Fig. 4.2d) [1–5].

Compression and expansion of the external ear represent two primary modes of structural transformation relevant to wearable device interaction. Compressive deformation is typically associated with cap-type and cup-type products, whereas expansive deformation arises at the tragus, cavum concha, cymba concha, and external acoustic meatus (EAM) due to ear-insert or plug-type designs (Figs. 4.1 and 4.2). Within these categories, distinct variables can be identified and classified according to their deformational behavior and perceptual significance.

To illustrate compressive processes, the application of a transparent polymethyl methacrylate (PMMA) plate with defined dimensions (e.g., 10 mm × 10 mm × 2 mm) can be used as a standardized means of exerting perpendicular pressure to the midsagittal plane. This approach allows the frontal contour deformation of the auricle to be recorded and analysed, providing a consistent framework for quantifying

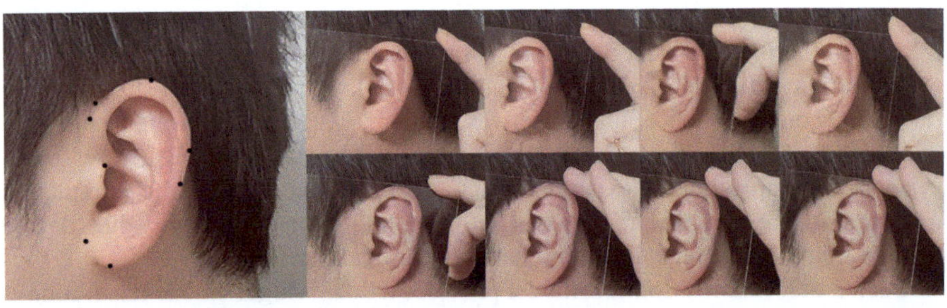

Fig. 4.1 Compression deformation of the auricle

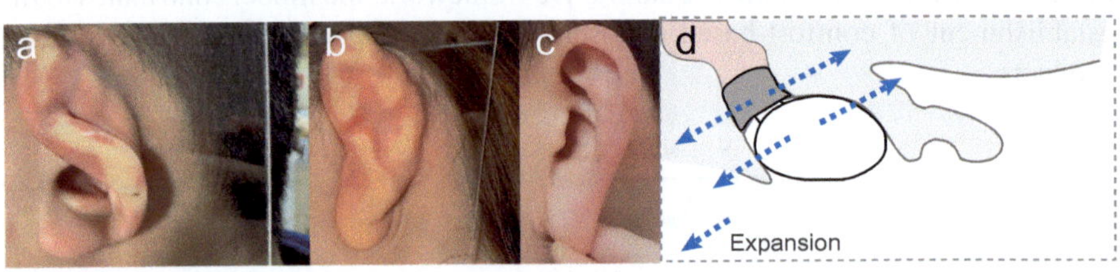

Fig. 4.2 Deformation of the ear protrusion (anterior and posterior), lobule, tragus, concha and external acoustic meatus: (**a**) compression (anterior ear protrusion); (**b**) compression (posterior ear protrusion); (**c**) tension (lobule); (**d**) expansion (tragus, concha and external acoustic meatus)

dimensional changes under controlled compression. Parameters such as auricular length (AL), auricular width at postaurale (AWP), auricular width at the incisura anterior auris posterior (AWI), and ear morphologic length (EML) consistently demonstrate sensitivity to compression and therefore constitute primary indicators of comfort-related deformation. In contrast, lobular length (LL) and lobular width (LW) display unusually high tolerance to stretching, often reaching nearly double the baseline size before discomfort becomes evident. This reflects the structural composition of the lobule and its dependence on localized external forces rather than intrinsic tissue deformation. For this reason, LL and LW are not considered reliable variables for comfort thresholding (Fig. 4.2c).

Auricular protrusion, both anterior and posterior, also presents particular characteristics. Even under conditions of full folding, deformation in these regions does not directly translate to discomfort at the auricular base. Hence, protrusion-related parameters are regarded as less critical for comfort assessment. Consistent with existing ergonomic principles [6, 7], conchal depth (CD) as well as other height, depth, and positional angle variables remain within the range of acceptable fit as long as detachment does not occur. For such cases, design practice generally allows a tolerance of approximately 5%, with additional validation achievable through dynamic actions such as head shaking or jumping. Consequently, CD is excluded from the critical parameter set.

Expansion of the EAM is subject to safety limitations in continuous testing. Therefore, circumference variables, entrance circumference (EC), first bend circumference (FBC), and second bend circumference (SBC), and area variables, entrance area (EA), first bend area (FBA), and second bend area (SBA), can be approximated indirectly through cross-sectional length and width, as these provide reliable control parameters for expansive deformation. The comfort-related thresholds of these variables can thus be estimated in practice by reference to the maximum expansive deformation ratio of either length or width.

Through above systematic differentiation, a set of auricular and EAM variables with direct relevance to wear comfort can be delineated. Parameters retained for thresholding include AL, AWP, AWI, and EML for compressive deformation, as well as conchal length (CL), conchal width (CW), cavum concha length (CCL), cavum concha width (CCW), tragal length (TL), entrance length (EL), entrance width (EW), first bend length (FBL), first bend width (FBW), second bend length (SBL), and second bend width (SBW) for dimensional adaptation and expansive deformation.

Demographic Composition

The demographics for external ear deformation and comfort characterization are established on the basis of health-related and age-specific considerations. Individuals aged 15–19 commonly present high prevalence of regular or continuous use of pain-suppressing medications, exceeding 70% over a 12-month period. Populations over 60 years of age likewise demonstrate more than 65% incidence of head-facial pain during the same timeframe (the samples come from Chap. 3). To avoid confounding effects on deformation ratios and comfort perception, both of these age groups are excluded from the analytical sample.

Within the selected range of 20–59 years, further control requirements are defined to ensure the reliability of external ear measurements. Participants are expected to abstain from strenuous physical activity for at least 1 week prior to involvement, reducing transient physiological influences on both the auricle and the external acoustic meatus (EAM). Exclusion principles additionally specify the absence of any head-facial discomfort within the preceding 12 months and no history of regular intake of pain-suppressing medications over the same period [8–11].

The resulting analytical sample comprises 320 individuals, evenly divided into 160 males and 160 females. To reflect developmental and age-related variation in external ear morphology and comfort perception, the cohort is stratified into four decades: twenties, thirties, forties, and fifties. Each decade contains 80 participants with an equal gender (1:1). This structural composition fulfills representativeness requirements and accords with the specifications provided in Table 2.2.

Comfort Data Collection

Within the context of external ear deformation, human comfort in wearable devices is defined as the perceptual response associated with structural compression, hereafter referred to as comfort perception. Comfort perception spans a continuum ranging from the absence of sensation to complete discomfort. For descriptive purposes, four reference states are distinguished: the absence of perceptual change (no discomfort), the sensation of skin compression (slight discomfort), the sensation of cartilage compression (moderate discomfort), and the sensation of pain (complete discomfort) [12, 13].

In applied assessments, slight discomfort, moderate discomfort, and complete discomfort represent the most salient perceptual categories, and the deformation ratios corresponding to these three values serve as critical measurement anchors [14]. To quantify these values, a comfort scaling system integrates the 5-point Likert scale and the Borg CR-10 scale [14–16]. Within this framework, comfort values are specifically defined as follows:

1. 0: no deformation (comfort, no discomfort),
2. −1: skin compression (slight discomfort),
3. −2: cartilage compression (moderate discomfort),
4. −5: pain (complete discomfort).

The methodological procedure is based on controlled deformation under standardized conditions. Reference points can be marked on the auricle to ensure geometric consistency during observation. Compression is applied perpendicularly to the sagittal plane of the cranium. For purposes of standardization, a transparent polymethyl methacrylate (PMMA) plate with dimensions of 10 mm × 10 mm × 2 mm can be employed to simulate progressive compression until the targeted perceptual levels are reached. The frontal view of the external ear is recorded at fixed distance, generating image sequences for subsequent analysis.

Temporal separation between trials is incorporated to prevent perceptual accumulation effects. At least 20 min between repeated compressions and no less than 24 h between bilateral measurements constitute the recommended intervals. For each identified comfort level, deformation ratios are determined by image-based geometric analysis using Eq. (4.1). To ensure data reliability, three measurements are obtained for each level, and the coefficient of variation (*CV*) is calculated. Only cases where $CV < 5\%$ are retained, and the mean value across the three trials is then computed to represent the effective comfort level for that participant.

Deformation-Comfort Modelling

The analysis of external ear deformation and comfort perception follows a structured framework combining statistical comparison and curve modelling. Differences across age groups, bilateral sides, and sex can be examined through one-way analysis of variance (ANOVA), paired-sample t-tests, and independent-sample t-tests, applied to deformation variables such as AL, AWP, AWI, and EML.

Curve fitting is introduced to capture the quantitative association between deformation ratios and comfort values. For auricular variables (AL, AWP, AWI, and EML), both individual and mean deformation ratios are modeled with reference to comfort values defined as 0, −1, −2, and −5. For tragus, conchal, and external acoustic meatus variables, corresponding deformation ratios align with the comfort anchors of 0% (no discomfort), 7% (slight discomfort), and 10% (complete discomfort), as outlined in Sect. 2.2.

Because comfort perception follows a monotonically decreasing trend, fitted curves are constrained to be either linear functions or monotonically decreasing segments of nonlinear functions. Exploratory fitting may include linear, exponential, logarithmic, power, logistic, and peak-type functions, while excluding overfitted solutions (using Origin 2019b, OriginLab Corp., US). Acceptance criteria incorporate adjusted R^2 values, comfort values ≤ 0, and the absence of local maxima or minima within the fitted interval.

To ensure robustness, comparative evaluations are conducted between individual-based and mean-based deformation ratio models. Comfort values are extended beyond the main anchors to intermediate levels, such as −0.5, −1.5, −2.5, −3.5, and −4.5, thereby forming a gradated spectrum from 0 (no discomfort) to −5 (complete discomfort). Each intermediate point is assigned to a corresponding deformation ratio, allowing continuous modelling across the full perceptual scale.

For enhanced resolution, the scale can be further discretized by generating 999 evenly spaced points between 0 and −5, yielding 1001 comfort values in total. Their associated deformation ratios are computed in both individual and mean models, followed by paired comparisons. Statistical significance is generally defined at a 95% confidence level ($\alpha = 0.05$), with results regarded as significant when $p < 0.05$.

4.1.3 Ethics Approval

The studies of this chapter performed in line with the principles of the Declaration of Helsinki. Informed consent was obtained from all participants. The studies of this chapter were approved by the Ethics Committee of Northwestern Polytechnic University (No. IRB No. 201502024, No. 202002024).

4.1.4 Model Fitting

Deformation-Comfort Model

For AL, AWP, AWI, and EML, deformation ratios show no significant effects of sex ($p = 0.10$–0.97), laterality ($p = 0.06$–0.97), or age-group distribution ($p = 0.06$–0.95). Detailed inferential outcomes are provided in Table C.1. Accordingly, data are pooled across sex, side, and age group for model fitting. Table 4.1 summarizes the pooled mean deformation ratios of AL, AWP, AWI, and EML at the corresponding comfort values.

For individual fittings of AL, AWP, AWI, and EML, the Logistic Exposure-Response Function (Logistic E-R, Eq. (4.2)) attains the highest goodness of fit. For pooled means, the First-order Exponential Decay Function (Period-intercept Version, Eq. (4.3)) performs best. For concha-, earlobe-, and EAM-related variables, the First-order Box-Lucas Function (Eq. (4.4)) provides the best fit.

For EAM circumference and area variables, wear-comfort values are recommended to be computed from the maximum expansive deformation ratio taken from either the corresponding length or width variable. Summary model outcomes appear

Table 4.1 Deformation ratios of auricular linear variables for wearing comfort

Variable	Comfort rating	Description	Ear, N	Mean (%)	SD
AL	−1	Slight discomfort	640	8.04	1.21
	−2	Moderate discomfort	640	11.51	0.91
	−5	Complete discomfort	640	16.03	1.28
AWP	−1	Slight discomfort	640	6.60	0.89
	−2	Moderate discomfort	640	11.03	1.28
	−5	Complete discomfort	640	16.69	1.36
AWI	−1	Slight discomfort	640	6.61	0.85
	−2	Moderate discomfort	640	12.41	1.30
	−5	Complete discomfort	640	16.80	1.22
EML	−1	Slight discomfort	640	5.70	0.93
	−2	Moderate discomfort	640	8.38	0.89
	−5	Complete discomfort	640	15.02	1.02

Table 4.2 Deformation-comfort curve fitting results

Variable	Data type of fitting	Deformation type	R-square	Adjusted R-square
AL	Individual	Compression	0.933	0.930
AL	Mean	Compression	0.9995	0.9993
EML	Individual	Compression	0.975	0.970
EML	Mean	Compression	0.998	0.995
AWP	Individual	Compression	0.944	0.941
AWP	Mean	Compression	0.998	0.996
AWI	Individual	Compression	0.952	0.951
AWI	Mean	Compression	0.989	0.969
CL, CW, CCL, CCW, TL, EL, EW, FBL, FBW, SBL, SBW	Reported data	Expansion	0.9999	0.9999

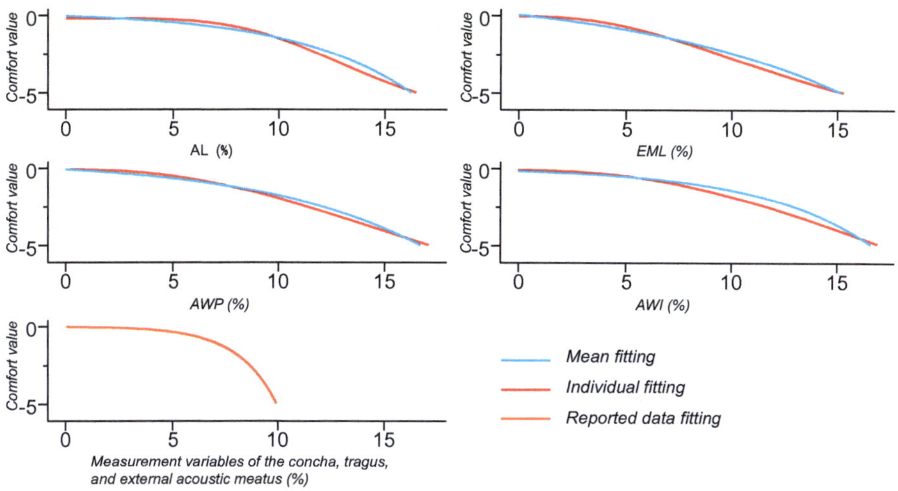

Fig. 4.3 Deformation-comfort curve fitting results

in Table 4.2; individual fittings are given in Eqs. (4.5)–(4.8), mean fittings in Eqs. (4.9)–(4.12), and fittings for concha, earlobe, and EAM-related variables in Eq. (4.13). Curve examples are shown in Fig. 4.3.

No significant overall differences are observed between individual and mean fittings at matched comfort values ($p > 0.05$). Accordingly, both data forms are admissible. Unless otherwise noted, threshold calculations adopt the mean fittings of AL, AWP, AWI, and EML to maintain consistency with the concha, earlobe, and EAM models. Individual threshold results for AL, AWP, AWI, and EML are listed in Tables C.2 and C.3.

$$y = \frac{A_1 - A_2}{1 + \left(\frac{x}{x_0}\right)^p} + A_2 \tag{4.2}$$

where, A_1 is the initial value of y; A_2 is the final value of y; x_0 is the centre point of the curve; p is the exponent.

$$y = y_0 + Ae^{-\frac{x}{t}} \tag{4.3}$$

where, $y_0$is the offset; A is the amplitude; t is the period intercept.

$$y = a\left(1 - e^{-bx}\right) \tag{4.4}$$

where, a is the coefficient, b is the coefficient.

$$g_{AL} = \frac{7.25661}{1 + \left(\frac{DR_{ext}}{14.12797}\right)^{4.38619}} - 7.42985,\ g_{AL} \in [0, -5] \tag{4.5}$$

$$g_{AWP} = \frac{9.21107}{1 + \left(\frac{DR_{ext}}{14.36459}\right)^{2.4017}} - 9.27213,\ g_{AWP} \in [0, -5] \tag{4.6}$$

$$g_{AWI} = \frac{13.57961}{1 + \left(\frac{DR_{ext}}{22.07721}\right)^{2.33817}} - 13.6609,\ g_{AWI} \in [0, -5] \tag{4.7}$$

$$g_{EML} = \frac{20.98124}{1 + \left(\frac{DR_{ext}}{29.3107}\right)^{2.26607}} - 21.067,\ g_{EML} \in [0, -5] \tag{4.8}$$

$$g_{AL} = 0.26602 - 0.28132e^{\frac{DR_{ext}}{5.47513}},\ g_{AL} \in [0, -5] \tag{4.9}$$

$$g_{AWP} = 0.59064 - 0.62994e^{\frac{DR_{ext}}{7.65429}},\ g_{AWP} \in [0, -5] \tag{4.10}$$

$$g_{AWI} = 0.06564 - 0.22762e^{\frac{DR_{ext}}{5.42836}},\ g_{AWI} \in [0, -5] \tag{4.11}$$

$$g_{EML} = 1.62950 - 1.59799e^{\frac{DR_{ext}}{10.64762}},\ g_{EML} \in [0, -5] \quad (4.12)$$

$$g_{con-eam} = 0.02511\left(1 - e^{0.52988DR_{expa}}\right),\ g_{con-eam} \in [0, -5] \quad (4.13)$$

where, g_{AL} is the comfort value reported by users regarding variations in auricle length; g_{AWP} is the comfort value reported by users regarding variations in the upper width of the auricle; g_{AWI} is the comfort value reported by users regarding variations in the lower width of the auricle; g_{EML} is the comfort value reported by users regarding variations in the basal length of the auricle; DR_{ext} is the deformation ratio caused by compression of the structure during product fitting; $g_{con-eam}$ is the comfort value reported by users regarding variations in measurement variables of the concha, tragus, and external acoustic meatus; DR_{expa} is the deformation ratio caused by expansion of the structure during product fitting.

Wearing Comfort Threshold and Comfort Perception Limit

Comfort values are defined at 0 (no discomfort), −0.5 (just-noticeable discomfort), −1 (slight discomfort), −1.5 (slight discomfort), −2 (moderate discomfort), −2.5 (moderate discomfort), −3 (high discomfort), −3.5 (high discomfort), −4 (extreme discomfort), −4.5 (extreme discomfort), and −5 (complete discomfort; comfort perception limit). Using Eqs. (4.9)–(4.13), deformation ratio-based wear-comfort thresholds and the comfort perception limit are computed. Results are summarized in Table 4.3.

Given that no significant differences in deformation ratios are detected across the existing age groups (20–59 years) for the individual variables, it is assumed (acknowledging a limitation) that the 15–19, 60–69, and 70–79 groups do not differ materially from the established range. On this basis, wearing comfort thresholds and comfort perception limits are computed from the mean measurement variables.

As summarized in Sect. 3.3, after controlling for height and weight, AL, AWP, AWI, EL, FBL, and SBL retain significant sex effects; EL also exhibits significant bilateral effects; and female means are smaller than left-right means. For universal, consumer-oriented design and scheme evaluation, it is therefore recommended to adopt the female values of AL, AWP, AWI, EL, FBL, and SBL for threshold calculations. For percentile- or age-specific applications, consult Appendix C. The resulting wear-comfort thresholds and comfort-perception limits based on mean variables for universal design and evaluation are reported in Table 4.4.

Table 4.3 Wearing comfort thresholds and comfort perception limits of deformation ratios (mean fitting)

Comfort rating	Description	Variable				
		AL (%)	AWP (%)	AWI (%)	EML (%)	CL, CW, CCL, CCW, TL, EL, EW, FBL, FBW, SBL, SBW (%)
0	No discomfort	0.00	0.00	0.00	0.00	0.00
−0.5	Just-noticeable discomfort	5.49	4.21	4.94	3.02	5.74
−1	Slight discomfort	8.23	7.10	8.37	5.26	7.01
−1.5	Slight discomfort	10.05	9.17	10.46	7.08	7.75
−2	Moderate discomfort	11.43	10.81	11.97	8.65	8.29
−2.5	Moderate discomfort	12.50	12.17	13.15	10.01	8.70
−3	High discomfort	13.42	13.32	14.11	11.22	9.05
−3.5	High discomfort	14.20	14.32	14.93	12.30	9.33
−4	Extreme discomfort	14.89	15.21	15.66	13.28	9.58
−4.5	Extreme discomfort	15.49	15.99	16.28	14.18	9.80
−5 (limit)	Complete discomfort	16.03	16.69	16.80	15.02	10.00
Type		Compressive	Compressive	Compressive	Compressive	Expansion

For design formulation and evaluation, mean-variable thresholds provide a practical, rapid reference. These can be tailored by incorporating characteristics of the target population, age, sex, and percentile, to obtain deformation-ratio-based thresholds and limits aligned with specific user groups. For evaluations aimed at individual users or small cohorts, it is recommended to compute parameters using the deformation-ratio wear-comfort thresholds and comfort-perception limits in conjunction with user-specific ergonomic measurements.

Table 4.4 Wearing comfort thresholds of deformation for mean size (mean fitting)

		Comfort rating										
Variable	Description	0	−0.5	−1	−1.5	−2	−2.5	−3	−3.5	−4	−4.5	−5 (limit)
AL (mm)	Female	64.92	68.48	70.26	71.44	72.34	73.04	73.63	74.14	74.59	74.98	75.33
AWP (mm)	Female	35.02	36.49	37.51	38.23	38.81	39.28	39.68	40.03	40.35	40.62	40.86
AWI (mm)	Female	27.52	28.88	29.82	30.40	30.81	31.14	31.40	31.63	31.83	32.00	32.14
EML (mm)	Total	56.43	58.13	59.40	60.43	61.31	62.08	62.76	63.37	63.92	64.43	64.91
CL (mm)	Total	29.46	31.15	31.53	31.74	31.90	32.02	32.13	32.21	32.28	32.35	32.41
CW (mm)	Total	19.62	20.75	21.00	21.14	21.25	21.33	21.40	21.45	21.50	21.54	21.58
CCL (mm)	Total	18.41	19.47	19.70	19.84	19.94	20.01	20.08	20.13	20.17	20.21	20.25
CCW (mm)	Total	15.37	16.25	16.45	16.56	16.64	16.71	16.76	16.80	16.84	16.88	16.91
TL (mm)	Total	18.25	19.30	19.53	19.66	19.76	19.84	19.90	19.95	20.00	20.04	20.08
EL (mm)	Female	15.00	15.86	16.05	16.16	16.24	16.31	16.36	16.40	16.44	16.47	16.50
EW (mm)	Total	8.60	9.09	9.20	9.27	9.31	9.35	9.38	9.40	9.42	9.44	9.46
FBL (mm)	Female	9.38	9.92	10.04	10.11	10.16	10.20	10.23	10.26	10.28	10.30	10.32
FBW (mm)	Total	7.70	8.14	8.24	8.30	8.34	8.37	8.40	8.42	8.44	8.45	8.47
SBL (mm)	Female	7.72	8.16	8.26	8.32	8.36	8.39	8.42	8.44	8.46	8.48	8.49
SBW (mm)	Total	7.25	7.67	7.76	7.81	7.85	7.88	7.91	7.93	7.94	7.96	7.98

4.2 Rapid Estimation Model for the Circumference and Area of the External Acoustic Meatus Cross-Sections

Circumference and area determine the contact perimeter and bearing surface for ear-insert elements. They are the natural denominators for pressure, stress, and deformation ratio mapping, enabling wear-comfort thresholds and comfort-perception limits to be set on a physically meaningful basis. Circumference governs geometric compatibility and leakage risk; area relates to contact stability under motion. Predicting these quantities supports sizing, tip geometry selection, and tolerance allocation for universal and percentile-targeted design. Cross-sectional areas and effective circumferences influence acoustic impedance, occlusion, and insertion loss. Predictive access to these measures facilitates trade-offs between comfort and acoustic targets. For expansive deformation at the entrance, first bend, and second bend, area provides the reference for distributed loads, while circumference indicates likely stress concentrations along the canal boundary, both critical for limiting local discomfort.

From an anatomical standpoint, the external acoustic meatus (EAM) is an internal, narrow, and curved canal embedded within the skull. This geometry, with bends, undercuts, and compliant soft-tissue boundaries, precludes rapid, direct acquisition of cross-sectional circumference and area. Visual access is inherently limited; contact instruments perturb the boundary through local deformation and orientation error; non-contact modalities introduce segmentation and calibration burdens, higher cost, and, for certain imaging methods, radiation considerations. Accordingly, perimeter and area are treated as derived rather than direct observables. A principled workaround is to model each EAM section (entrance, first bend, second bend) as near-elliptical and to adopt readily measurable linear primitives, length-width pairs (EL/EW, FBL/FBW, SBL/SBW), as the major-minor axes. Closed-form estimators (e.g., Srinivasan-Ramanujan ellipse formulations) then yield circumference and area with high computational efficiency and acceptable fidelity for design inference. This theoretical stance preserves speed, repeatability, and cross-study comparability, while supplying the derived quantities required for comfort-threshold mapping, fit and sealing evaluation, load distribution analysis, and universal or percentile-targeted product specifications. In short, anatomical inaccessibility converts a measurement problem into a modelling problem, for which ellipse-based prediction provides a coherent and ergonomically meaningful solution.

4.2.1 Methods of Rapid Estimation

Predictive analysis of the EAM is framed as a model-to-measure pipeline in which functional forms are selected only when age-morphology associations are supported by evidence. Two complementary trajectories are considered: persistent age-growth (continuous, gradual change) and phasic age-growth (stage-dominant change). Accordingly, linear or constrained nonlinear fittings are adopted contingent on the outcomes of age-group difference analysis and growth-trajectory analysis. For the entrance, first bend, and second bend, the length-width pairs (EL/EW, FBL/FBW, SBL/SBW) serve as the major-minor axes of a near-elliptical section, from which area (Eq. (4.14)) and circumference (Eq. (4.15)) are derived. Eq. (4.15) follows the Srinivasan-Ramanujan ellipse formulation [17], providing closed-form estimates that are computationally efficient and measurement-aligned. Predictive fidelity is verified with regression R^2 at three scales, individual, age-specific mean, and age-group mean, ensuring coherence across analytic levels and supporting subsequent threshold setting and design evaluation.

$$A_{ellipse} = \pi ab \tag{4.14}$$

$$C_{ellips} = \pi(a+b)\left(1+\frac{3h}{10+\sqrt{4-3h}}\right) \tag{4.15}$$

where, $h=\frac{(a-b)^2}{(a+b)^2}$, $a=\frac{Major\ Axis}{2}$, $b=\frac{Minor\ Axis}{2}$, $\pi = 3.14159$; $A_{ellipse}$ is the estimated cross-sectional area; $C_{ellipse}$ is the estimated cross-sectional circumference.

As summarized in Table 4.5, ellipse-based estimates treat EL/EW, FBL/FBW, and SBL/SBW as the major-minor axes for the entrance, first bend, and second bend, respectively, from which cross-sectional area and circumference are derived (Eqs. (4.14) and (4.15)). The estimates show very high concordance with measured values or means ($R = 0.901$–0.999; $R^2 = 0.811$–0.999). Because EL/EW, FBL/FBW, and SBL/SBW can be obtained directly with a Vernier caliper via molding, whereas perimeter and area cannot, the ellipse outputs provide practical approximations when ultra-high precision is unnecessary. For applications demanding tighter accuracy, Table 4.5 reports rapid estimation models (section-specific linear regressions) for circumference and area; the corresponding formulas are given in Eqs. (4.16)–(4.33). An example of model fidelity is illustrated by the linear relationship between estimated and measured second-bend area in Fig. 4.4.

4.2.2 Rapid Estimation Models of the External Acoustic Meatus Cross-Sectional Circumference and Area

Rapid Estimation Models Using Individual-Level Data

$$EC = 0.87 + 1.02\left(\frac{\pi(EL+EW)}{2}\left(1+\frac{\frac{(EL-EW)^2}{(EL+EW)^2}}{10+\sqrt{4-\frac{3(EL-EW)^2}{(EL+EW)^2}}}\right)\right) \tag{4.16}$$

Table 4.5 Rapid estimation models of cross-sections for the external acoustic meatus

Independent variable	Dependent variable	Long axis	Short axis	Individual values (Ear, N = 2240)					
(estimated)	(measured)	(measured)	(measured)	R	R-square	Adjusted R-square	Intercept	Coefficient	Standard error of estimate
Circumference	EC (mm)	EL	EW	0.904	0.818	0.818	0.87*	1.02**	1.22
Area	EA (mm^2)	EL	EW	0.959	0.920	0.920	2.39**	0.98**	4.44
Circumference	FBC (mm)	FBL	FBW	0.954	0.910	0.910	0.86**	0.99**	0.84
Area	FBA (mm^2)	FBL	FBW	0.989	0.978	0.978	1.22**	0.97**	1.66
Circumference	SBC (mm)	SBL	SBW	0.901	0.811	0.811	0.83**	0.97**	0.90
Area	SBA (mm^2)	SBL	SBW	0.961	0.923	0.923	1.16**	0.97**	2.04
Independent variable	Dependent variable	Long axis	Short axis	Age-based mean data (age, N = 65)					
(estimated)	(measured)	(measured)	(measured)	R	R-square	Adjusted R-square	Intercept	Coefficient	Standard error of estimate
Circumference	EC (mm)	EL	EW	0.953	0.908	0.906	0.80*	1.02**	0.63
Area	EA (mm^2)	EL	EW	0.983	0.966	0.966	1.12*	0.99**	1.70
Circumference	FBC (mm)	FBL	FBW	0.983	0.967	0.966	0.53**	1.01**	0.32
Area	FBA (mm^2)	FBL	FBW	0.995	0.989	0.989	2.92**	0.94**	0.67
Circumference	SBC (mm)	SBL	SBW	0.951	0.905	0.904	2.86**	0.89**	0.28
Area	SBA (mm^2)	SBL	SBW	0.985	0.971	0.971	1.47**	0.96**	0.62
Independent variable	Dependent variable	Long axis	Short axis	Age-group-based mean data (age group, N = 7)					
(estimated)	(measured)	(measured)	(measured)	R	R-square	Adjusted R-square	Intercept	Coefficient	Standard error of estimate
Circumference	EC (mm)	EL	EW	0.971	0.943	0.932	2.43	0.98**	0.56
Area	EA (mm^2)	EL	EW	0.998	0.996	0.995	−0.92	1.01**	0.73

(estimated)	(measured)	(measured)	(measured)	R	R-square	Adjusted R-square	Intercept	Coefficient	Standard error of estimate
Circumference	FBC (mm)	FBL	FBW	0.998	0.996	0.995	−0.07	1.02**	0.13
Area	FBA (mm^2)	FBL	FBW	0.999	0.999	0.999	2.01*	0.96**	0.23
Circumference	SBC (mm)	SBL	SBW	0.992	0.984	0.981	2.60	0.90**	0.13
Area	SBA (mm^2)	SBL	SBW	0.998	0.995	0.994	1.67	0.96**	0.30

**$p < 0.01$, *$p < 0.05$

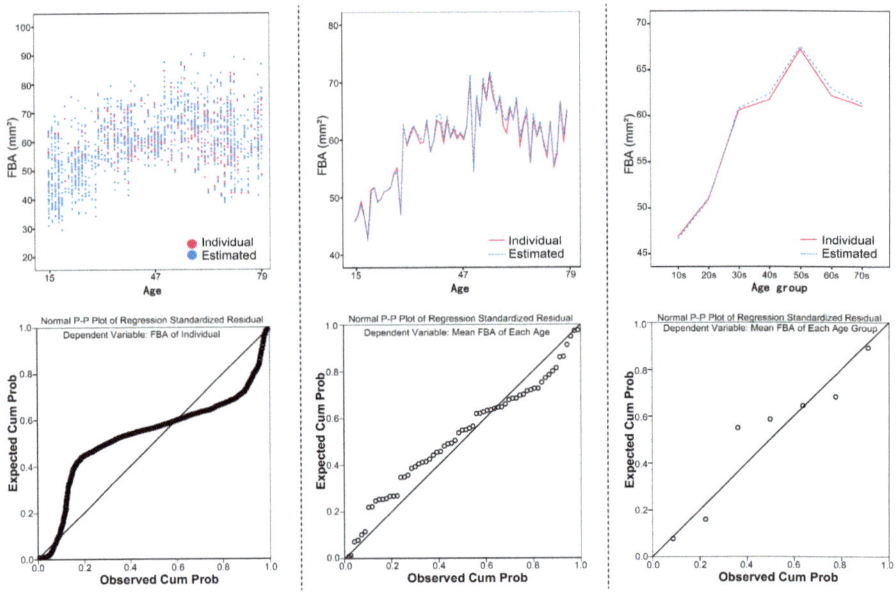

Fig. 4.4 Linear regression results of dimension rapid estimation for the second bend area

$$EA = 2.39 + 0.98\left(\frac{\pi\left(EL \times EW\right)}{4}\right) \tag{4.17}$$

$$FBC = 0.86 + 0.99\left(\frac{\pi\left(FBL + FBW\right)}{2}\left(1 + \frac{\frac{\left(FBL - FBW\right)^2}{\left(FBL + FBW\right)^2}}{10 + \sqrt{4 - \frac{3\left(FBL - FBW\right)^2}{\left(FBL + FBW\right)^2}}}\right)\right) \tag{4.18}$$

$$FBA = 1.22 + 0.97\left(\frac{\pi\left(FBL \times FBW\right)}{4}\right) \tag{4.19}$$

$$SBC = 0.83 + 0.97\left(\frac{\pi\left(SBL + SBW\right)}{2}\left(1 + \frac{\frac{\left(SBL - SBW\right)^2}{\left(SBL + SBW\right)^2}}{10 + \sqrt{4 - \frac{3\left(SBL - SBW\right)^2}{\left(SBL + SBW\right)^2}}}\right)\right) \tag{4.20}$$

$$SBA = 1.16 + 0.97\left(\frac{\pi(SBL \times SBW)}{4}\right) \tag{4.21}$$

Rapid Estimation Models Using Age-Based Mean Data

$$EC = 0.80 + 1.02\left(\frac{\pi(EL + EW)}{2}\left(1 + \frac{\frac{(EL - EW)^2}{(EL + EW)^2}}{10 + \sqrt{4 - \frac{3(EL - EW)^2}{(EL + EW)^2}}}\right)\right) \tag{4.22}$$

$$EA = 1.12 + 0.99\left(\frac{\pi(EL \times EW)}{4}\right) \tag{4.23}$$

$$FBC = 0.53 + 1.01\left(\frac{\pi(FBL + FBW)}{2}\left(1 + \frac{\frac{(FBL - FBW)^2}{(FBL + FBW)^2}}{10 + \sqrt{4 - \frac{3(FBL - FBW)^2}{(FBL + FBW)^2}}}\right)\right) \tag{4.24}$$

$$FBA = 2.92 + 0.94\left(\frac{\pi(FBL \times FBW)}{4}\right) \tag{4.25}$$

$$SBC = 2.86 + 0.89\left(\frac{\pi(SBL + SBW)}{2}\left(1 + \frac{\frac{(SBL - SBW)^2}{(SBL + SBW)^2}}{10 + \sqrt{4 - \frac{3(SBL - SBW)^2}{(SBL + SBW)^2}}}\right)\right) \tag{4.26}$$

$$SBA = 1.47 + 0.96\left(\frac{\pi(SBL \times SBW)}{4}\right) \tag{4.27}$$

Rapid Estimation Models Using Age-Group-Based Mean Data

$$EC = 2.43 + 0.89\left(\frac{\pi\left(EL+EW\right)}{2}\left(1+\frac{\frac{\left(EL-EW\right)^2}{\left(EL+EW\right)^2}}{10+\sqrt{4-\frac{3\left(EL-EW\right)^2}{\left(EL+EW\right)^2}}}\right)\right) \tag{4.28}$$

$$EA = -0.92 + 1.01\left(\frac{\pi\left(EL \times EW\right)}{4}\right) \tag{4.29}$$

$$FBC = -0.07 + 1.02\left(\frac{\pi\left(FBL+FBW\right)}{2}\left(1+\frac{\frac{\left(FBL-FBW\right)^2}{\left(FBL+FBW\right)^2}}{10+\sqrt{4-\frac{3\left(FBL-FBW\right)^2}{\left(FBL+FBW\right)^2}}}\right)\right) \tag{4.30}$$

$$FBA = 2.01 + 0.96\left(\frac{\pi\left(FBL \times FBW\right)}{4}\right) \tag{4.31}$$

$$SBC = 2.60 + 0.90\left(\frac{\pi\left(SBL+SBW\right)}{2}\left(1+\frac{\frac{\left(SBL-SBW\right)^2}{\left(SBL+SBW\right)^2}}{10+\sqrt{4-\frac{3\left(SBL-SBW\right)^2}{\left(SBL+SBW\right)^2}}}\right)\right) \tag{4.32}$$

$$SBA = 1.67 + 0.96\left(\frac{\pi\left(SBL \times SBW\right)}{4}\right) \tag{4.33}$$

4.3 Phasic Age-Growth Models and Persistent Age-Growth Models of Measurement Variables

Universal design for users with special needs, such as occupation specific industrial equipment or assistive devices for individuals with hearing impairment, typically targets products with long service lives [18–20]. Under such horizons, design validation requires a quantitative mapping from age to external ear morphology (auricular and, where appropriate, EAM) and to the associated comfort indices based on deformation ratio. A predictive relationship is therefore formulated to express dimensional change as a function of age, with trajectories interpreted through persistent age-growth for continuous gradual change and phasic age-growth for stage-dominant change. This mapping enables forward estimation of target users' morphological dimensions and derived wear comfort values at future ages, providing a principled basis for checking whether current parameter choices remain within comfort thresholds and perception limits across the intended service life. To preserve parsimony and interpretability, potential collinearity among candidate predictors is examined prior to fitting. Variance inflation factor diagnostics for age, height, weight, and BMI fall in the range 5.21–6.00, indicating potential or non-severe multicollinearity. Accordingly, age is retained as the sole independent variable. The resulting specification supports lifetime compliance checks, percentile targeting, and iterative scheme evaluation within a universal design context.

4.3.1 Model Fitting

Sections 3.3 and 3.4 indicate that AIA, TH, and CD show no significant association with age and present waveform like fluctuations with high between person variability. These three variables are excluded. Section 4.2 shows that EC, EA, FBC, FBA, SBC, and SBA can be obtained by rapid estimation and are not modeled directly here. For the remaining 32 measurement variables, two pattern classes of age growth curves are identified:

1. Monotonically increasing curves: 22 persistent age-growth variables: AL, AWP, AWI, CL, EML, LL, LW, CW, CCL, CCW, TL, EL, SBL, SD-I, SD-II-F, SD-II-E, ED-P, ED-C, ACE, AS-I, ES-I, AS-II.
2. Quadratic polynomial or bell-shaped curves: Nine phasic age-growth variables: EPS, EPT, CAP, CAO, EW, FBL, FBW, SBW, ES-II.

Accordingly, polynomial or unimodal fitting is applied to the nine phasic age growth variables, while batch curve exploration is applied to the 22 persistent age growth variables under the constraint that comfort-related trends remain monotonic where applicable. Overfitted solutions are rejected. Final model selection favors the specification that achieves the best fitting performance and aligns with the observed trends of age group means.

4.3.2 Models of Measurement Variables

Persistent Age-Growth Measurement Variables

Among the 22 persistent age growth measurement variables, CW, CCL, CCW, TL, and SBL achieve the highest goodness of fit with the First-order Box-Lucas Function (Eq. (4.3); adjusted-R^2 between 0.767 and 0.959). AWP and AS-II fit best with the Logistic E-R Function (Eq. (4.2); adjusted-R^2 between 0.774 and 0.819). AL, LL, EML, EL, ED-I, ED-II-E, ED-P, ACE, AS-I, and ES-I are best captured by the Two-parameter Logarithmic Form (Eq. (4.34); adjusted-R^2 between 0.782 and 0.973), with AL showing the highest adjusted-R^2. ED-C and ED-II-F are best modeled by the S Logistic function (Eq. (4.35); adjusted-R^2 between 0.739 and 0.837). AWI, CL, and LW are best modeled by the Farazdaghi-Harris Function (Eq. (4.36); adjusted R^2 between 0.771 and 0.894). The quantitative relationships for all 22 variables with age are provided in Eqs. (4.37)–(4.58).

$$y = b\ln\left(x - a\right) \tag{4.34}$$

where, a is the offset; b is the coefficient.

$$y = \frac{a}{1 + e^{-k\left(x - x_c\right)}} \tag{4.35}$$

where, a is the amplitude; x_c is the centre point; k is the coefficient.

$$y = \left(a + bx^c\right)^{-1} \tag{4.36}$$

where, a is the coefficient; b is the coefficient; c is the exponent

$$CW = 20.01706\left(1 - e^{-0.14125Age}\right) \tag{4.37}$$

$$CCL = 18.80126\left(1 - e^{-0.15717Age}\right) \tag{4.38}$$

$$CCW = 15.6991\left(1 - e^{-0.13493Age}\right) \tag{4.39}$$

$$TL = 18.64155\left(1 - e^{-0.1571Age}\right) \tag{4.40}$$

$$SBL = 8.00037\left(1 - e^{-0.17248Age}\right) \tag{4.41}$$

$$AWP = 7331.14099 - \frac{7298.71611}{1 + \left(\frac{Age}{8187.32162}\right)^{1.48604}} \tag{4.42}$$

$$AS-II=139.67618-\frac{7.15925}{1+\left(\frac{Age}{66.98212}\right)^{10.06074}} \tag{4.43}$$

$$AL=15.43427\ln\left(Age+31.09349\right) \tag{4.44}$$

$$LL=4.89525\ln\left(Age+2.94708\right) \tag{4.45}$$

$$EML=12.99577\ln\left(Age+36.21699\right) \tag{4.46}$$

$$EL=3.54192\ln\left(Age+35.26739\right) \tag{4.47}$$

$$ED-I=1.93712\ln\left(Age+7.40071\right) \tag{4.48}$$

$$ED-\Pi-E=3.37281\ln\left(Age+14.40798\right) \tag{4.49}$$

$$ED-P=2.65127\ln\left(Age+22.78835\right) \tag{4.50}$$

$$ACE=6.10693\ln\left(Age+17.38197\right) \tag{4.51}$$

$$AS-I=20.57897\ln\left(Age+150.69239\right) \tag{4.52}$$

$$ES-I=4.16001\ln\left(Age+0.29733\right) \tag{4.53}$$

$$ED-C=\frac{15.03788}{1+e^{-0.06359\left(Age+9.12043\right)}} \tag{4.54}$$

$$ED-II-F=\frac{9.63176}{1+e^{-0.01273\left(Age+43.68859\right)}} \tag{4.55}$$

$$AWI=\left(0.03874-0.00007Age^{1.03251}\right)^{-1} \tag{4.56}$$

$$CL=\left(0.05625-0.000001Age^{1.50352}\right)^{-1} \tag{4.57}$$

$$LW=\left(0.03622-0.00001Age^{1.90563}\right)^{-1} \tag{4.58}$$

where, $Age \in [15,79]$.

Phasic Age-Growth Measurement

Among the nine phasic age-growth variables, EPS, EPT, CAP, CAO, FBL, and ES-II are best modeled by the Lorentzian Unimodal Function (Eq. (4.59); adjusted $R^2 = 0.836$–0.908). EW, FBW, and SBW are best modeled by the quadratic polynomial (Eq. (4.60); adjusted $R^2 = 0.901 \sim 0.949$). The age-measurement relationships for all nine variables are given in Eqs. (4.61)–(4.69), and representative fitted curves for each function are shown in Fig. 4.5.

$$y = y_0 + \left(\frac{2A}{\pi}\right)\left(\frac{w}{4(x - x_c)^2 + w^2}\right) \tag{4.59}$$

where, y_0 is the offset; A is the Area; x_c is the centre point; w is the full width at half maximum.

$$y = A + Bx + Cx^2 \tag{4.60}$$

where, A is the Intercept; B is the coefficient; C is the coefficient.

$$EPS = 12.15791 + \frac{220639.60465}{4\pi(Age - 49.15235)^2 + 13855.09087\pi} \tag{4.61}$$

$$EPT = 22.99902 + \frac{5800.13674}{4\pi(Age - 45.42177)^2 + 893.51492\pi} \tag{4.62}$$

$$CAP = 33.08362 + \frac{210114.07617}{4\pi(Age - 34.47746)^2 + 7195.8795\pi} \tag{4.63}$$

$$CAO = 53.18825 + \frac{28800.546475}{4\pi(Age - 32.65415)^2 + 1183.87468\pi} \tag{4.64}$$

$$FBL = 4.54315 + \frac{256818.80068}{4\pi(Age - 60.38542)^2 + 13724.00535\pi} \tag{4.65}$$

$$ES - II = 24.96022 + \frac{40782.03567}{4\pi(Age - 53.15081)^2 + 1358.39938\pi} \tag{4.66}$$

$$EW = 6.83226 + 0.08792Age - 0.0009Age^2 \tag{4.67}$$

$$FBW = 6.20541 + 0.06904Age - 0.00067Age^2 \tag{4.68}$$

$$SBW = 5.17155 + 0.09238Age - 0.00083Age^2 \tag{4.69}$$

where, $Age \in [15, 79]$.

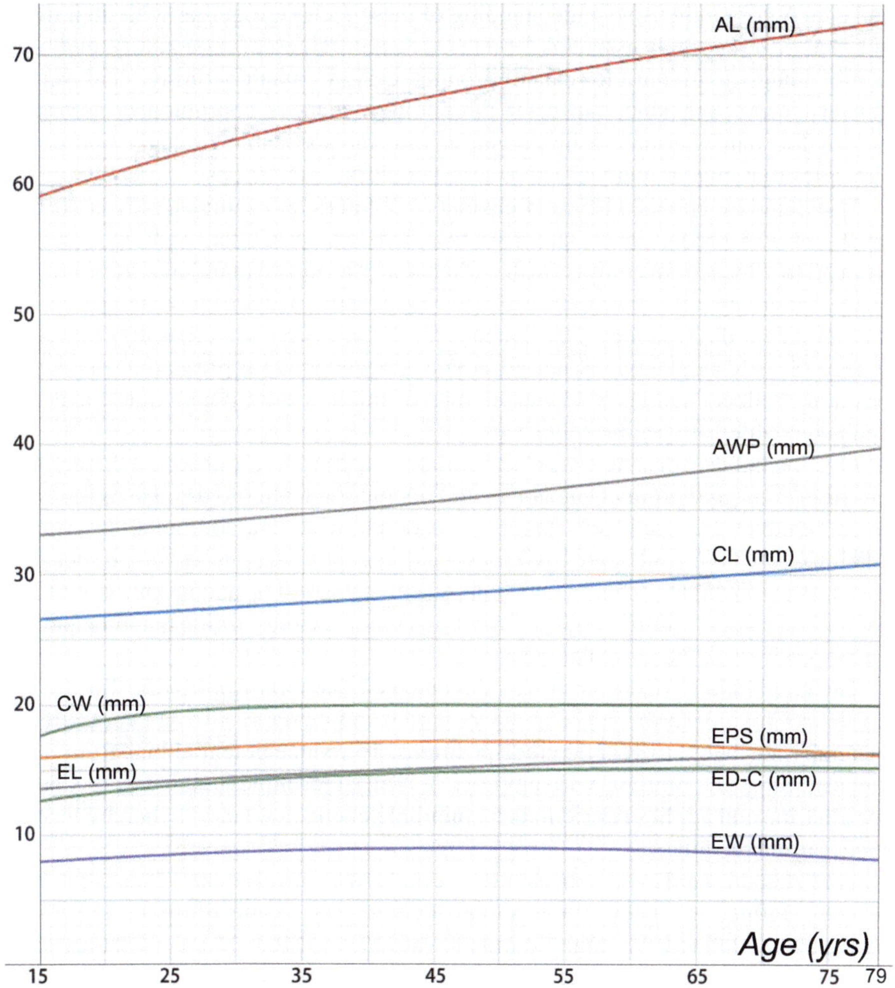

Fig. 4.5 Age-variable curve fitting results

4.4 Morphology, Growth, and Comfort

4.4.1 Auricle: Population Structure, Measures, and Growth Dynamics

The auricle constitutes both a biological identifier of the craniofacial system and the first structural element in the regulation of auditory input. Its morphology is embedded within cultural, social, physiological, psychological, ergonomic, and cognitive contexts, which explains the growing scholarly and design interest in its systematic characterization. Dimensional features of the auricle vary with ethnicity, sex, side, and age. A comprehensive understanding of these factors, together with their distributional differences, is essential for achieving naturalness, functionality, and ergonomic integrity in auricle-related product design, manufacturing, and service delivery. Moreover, since auricular morphology changes continuously across the human lifespan, age-linked investigations are indispensable for universal design strategies that must remain effective across multiple age groups.

The database covers 1120 participants (2240 auricles) sampled across 30 first-level and 133 second-level administrative divisions in China, with ages ranging from 15 to 79 years. In contrast with earlier work based on small laboratory samples (≤15 individuals) or convenience samples (≥30 individuals from a subgroup), this dataset meets the baseline requirements for constructing a morphology reference library [21]. Nonetheless, as not all ethnic subgroups are represented, the dataset primarily reflects aggregate characteristics of the domestic population rather than detailed profiles of specific subgroups.

Over recent decades, new auricular variables appear in response to demands arising from emerging product categories and interaction technologies. Historically, 13 fixed variables are used; this framework extends with five additional descriptors, AWI, CCL, CCW, CAP, and CAO, designed to capture cavum concha dimensions and protrusion angles. These extensions enable coverage of unconventional fittings for ear-cup and ear-muff products, cavum-specific earplug fittings, and headband-based products at the auricular base.

Geographic heterogeneity is well documented. Reports from India, Turkey, and northern China describe regional asymmetries and deviations from overall trends, generally attributed to genetic predispositions. To reduce potential bias from fixed-site sampling, the present work employs hub-city recruitment across both sides of the Hu Line and applies restrictions on familial genetic association, thereby increasing representativeness.

Age-related analyses consistently reveal two phenomena: Persistent age-growth, defined as steady dimensional increase across the lifespan, and Phasic age-growth, defined as stage-dominant variability. Most linear variables show persistent increases with age; a distinctive feature of the auricle compared with other organs. This property underscores the necessity of periodically revising universal design strategies, particularly for adolescent and elderly populations. Eleven linear variables function as persistent age-growth indices: AL, AWP, AWI, CL, EML, LL, LW, CW, CCL, CCW, and TL. Similarities in their growth patterns are further supported by factor analysis, which validates the use of FASVs (factors associated with shape variations) to represent auricular growth trajectories. By contrast, AIA, TH, and CD show no significant age associations. AIA, in particular, exhibits high inter-individual variability independent of age, height, weight, or BMI, suggesting that mean-based adjustment or exclusion is preferable in design applications.

Among the persistent variables, LL shows the greatest age-linked increase (~32%), consistent with findings from previous studies. To control artifacts induced by earring use, participants with pierced ears are excluded; the observed growth is therefore attributed to gradual loss of elasticity in lobular skin lacking cartilage. Unlike findings in Iranian, Indian, and Sudanese Arab populations, no significant lobular asymmetry is observed in the present Chinese sample ($p > 0.05$). These results are directly relevant to the design of ear-cup and ear-muff products covering the lobule, to domestic health-monitoring devices using lobular tissue (e.g., for blood glucose detection), and to cross-population comparative analyses.

Auricular protrusions (EPS, EPT, CAP, CAO) hold critical roles in both clinical diagnosis and design practice. Evidence shows that full auricular folding does not induce discomfort at the auricular base, implying that protrusion values may be treated as comfortable unless detachment occurs, or approximated using 5% variability. An additional observation is that protrusion levels at ages 15–19 and 70–79 are comparable, a trend not previously reported. Earlier findings suggest that protrusion growth may reflect changes in cavum depth or antihelix extension. Since cavum depth shows no age-related trend in this dataset, the latter explanation is more consistent with domestic observations.

Bilateral asymmetry functions as an important design consideration for paired products. Although some northern Chinese groups display significant side differences, no overall asymmetry is evident in the nationwide sample, indicating that asymmetry is regional rather than universal. Representative human templates may therefore require regional adjustments for optimal ergonomic validity. Sexual dimorphism also remains a fundamental determinant. In universal design practice, dimensions from the sex with smaller values ($p < 0.05$) are often adopted as the safer reference. Consistent with Table 3.10, female data are recommended for AL, AWP, AWI, EPS, EPT, and CAP, with overall means applicable to the remainder.

Unlike most auricular dimensions, the deformation ratios of AL, AWP, AWI, and EML under compression show no significant variation across age, sex, or side. Mean-fitted and aggregated individual-fitted models yield comparable outcomes at equivalent comfort values, and both forms are therefore acceptable. Analyses based on age-group means indicate that persistent age-growth variables follow exponential or logarithmic trajectories rather than linear ones, refining assumptions made in earlier studies.

4.4.2 External Acoustic Meatus: Measurement Schema, Age Dynamics, and Design Translation

The external acoustic meatus (EAM) is an indispensable component of the auditory pathway and a primary site of interaction for in-ear devices. Despite its functional importance, comprehensive and standardized characterization of the canal remains limited. A coherent measurement framework includes linear, angular, circumference, and area variables, together with spatial descriptors of entrance and isthmus positions relative to auricular and craniofacial landmarks.

A 22-variable system is used to capture the morphology of the EAM across three cross-sections: entrance, first bend, and second bend. This system includes lengths, widths, circumferences, and areas (variables 19–30); depth measures of canal segments (variables 31–33); entrance-positioning variables relative to the cavum midpoint and auricular plane (variables 34–36); and orientation-angle variables describing isthmus position relative to cranial planes (variables 37–40). With these descriptors, the morphology of 1120 participants is quantified, producing a population-level reference dataset for ergonomic and design applications.

Cross-sectional analysis shows significant age-linked transitions at three stages: from 15–19 to 20–29 years (19 variables), from 20–29 to 30–39 years (19 variables), and from 50–59 to 60–69 years (15 variables) (Table 3.9). EL increases overall by 20.20%, comparable to auricular length (AL, 20.02%). By contrast, EW declines markedly in older groups, with reductions evident from the age of 60 onward. Section-specific patterns indicate that the first-bend area decreases progressively after age 49, while the second-bend area decreases from age 39. The age trajectory of the EAM can therefore be described as a gradual enlargement of the entrance accompanied by midlife reductions in the areas of the internal bends, a pattern that aligns with anatomical explanations of age-related hearing loss.

Entrance-positioning variables (ED-C, ED-P, ACE) follow trajectories similar to the cavum midpoint, with lateral and posterior displacement relative to the Frankfurt plane and the craniofacial sagittal plane. These changes are attributed to skin thickening and the progressive stretching of tragal and antihelix cartilage with age. The result is a lateral rotation of the canal entrance around a vertical axis through its posterior landmark and perpendicular to the auricular plane. This transformation has direct implications for in-ear device stability, particularly for elderly users.

Depth and orientation-angle variables of the isthmus increase steadily with age, indicating progressive deepening, partial straightening, and posterior-superior displacement of the intracranial canal segment. To capture these dynamics, seven FASVs (factors associated with shape variations) are introduced for the canal.

Sexual dimorphism is evident in most variables, and 12 remain significant after controlling for height and weight. For universal comfort design, female values are recommended for EL, EC, EA, FBL, FBC, FBA, SBL, SD-II-F, ED-C, ED-P, AS-I, and AS-II, while overall means are appropriate for the remainder. Bilateral asymmetry is negligible overall, with the exception of EL and EC. For paired in-ear device design, slightly larger left-ear reference values may improve balance, though further validation is necessary. A tentative explanation attributes this asymmetry to habitual unilateral device use: 64% of participants report regular single-ear use, with 73% of these favouring the left ear. Prolonged unilateral loading may therefore contribute to cartilage expansion at the entrance.

Cross-sectional geometry is confirmed to be near-elliptical both visually and numerically across age groups. Elliptical approximation is thus suitable for rapid estimation of circumference and area. Since lengths and widths can be directly measured from casts, ellipse-based predictions provide practical surrogates when ultra-high precision is not required. Regression-based estimation further improves accuracy without the need for 3D scanning.

4.4.3 Typology of Variables and Comfort Modelling

Integration of age-group distributions with factor analysis enables the classification of the external-ear measurement system into three functional categories: persistent age-growth variables, phasic age-growth variables, and non-age-related variables. This typology provides a structured basis for linking morphology to comfort modelling.

Persistent age-growth variables exhibit monotonic increases with age and therefore serve as robust predictors for long-term dimensional trends. They are directly applicable to universal design scenarios in which products must accommodate gradual, continuous change throughout the service life. Phasic age-growth variables demonstrate transitional behavior, including growth, stabilization, and decline phases, and thereby identify critical age windows where fit and comfort may be at risk. Non-age-related variables, exemplified by AIA, TH, and CD, remain stable across age and provide invariant reference points for product design across demographic groups. This classification informs the construction of deformation-comfort models. Compression models for outer-contour variables (AL, AWP, AWI, EML) exhibit stability across age, sex, and side, supporting the definition of universal compression thresholds in terms of deformation ratio. By contrast, expansion models for concha- and EAM-related variables (CL, CW, CCL, CCW, TL, EL, EW, FBL, FBW, SBL, SBW) reveal pronounced variability across sex and age. These patterns demonstrate the limitations of single mean values for design purposes and emphasize the necessity of adopting percentile-based criteria when defining thresholds. Within this framework, design policies can be stratified according to intended scope. Universal policies rely on female reference values where sexual dimorphism is significant, ensuring coverage for the broadest user base. Regionalized policies adjust for geographic or subgroup variation, improving naturalness and cultural alignment. Targeted policies focus on specific cohorts, defined by age, sex, or percentile, to ensure that product dimensions and comfort limits are responsive to distinct needs.

Competing Interests This study is by "the Fundamental Research Funds for the Central Universities" (NPU Grant No. D5000250035, SEU Grant No. 2242024S30013). Additional support comes from the Northwestern Polytechnical University. The authors have no conflicts of interest to declare that are relevant to the content of this chapter.

References

1. Fan, H., Yu, S., Chu, J., et al. (2019). Anthropometric characteristics and product categorization of Chinese auricles for ergonomic design. *International Journal of Industrial Ergonomics, 69*, 118–141. https://doi.org/10.1016/j.ergon.2018.11.002
2. Fan, H., Yu, S., Wang, M., et al. (2021). Analysis of the external acoustic meatus for ergonomic design: Part I - Measurement of the external acoustic meatus using casting, scanning and rapid estimation approaches. *Ergonomics, 64*(5), 640–656. https://doi.org/10.1080/00140139.2020.1858188
3. Fan, H., Yu, S., Wang, M., et al. (2021). Analysis of the external acoustic meatus for ergonomic design: Part II - Anthropometric variations of the external acoustic meatus by sex, age and side in Chinese population. *Ergonomics, 64*(5), 657–670. https://doi.org/10.1080/00140139.2020.1867769
4. Wang, M., Fan, H., Harris, C., et al. (2019). Anthropometric characteristics of Chinese auricles for ergonomic design. *Proceedings of the Human Factors and Ergonomics Society Annual Meeting, 63*(1), 1962–1963. https://doi.org/10.1177/1071181319631275
5. Fan, H., Wang, M., Adamson, C. H., et al. (2020). Anthropometric analysis of external acoustic meatus for ergonomic design. *Proceedings of the Human Factors and Ergonomics Society Annual Meeting, 64*(1), 1833–1835. https://doi.org/10.1177/1071181320641397
6. Doutres, O., Sgard, F., Terroir, J., Perrin, N., Jolly, C., Gauvin, C., & Negrini, A. (2019). A critical review of the literature on comfort of hearing protection devices: Definition of comfort and identification of its main attributes for earplug types. *International Journal of Audiology, 58*(12), 824–833. https://doi.org/10.1080/14992027.2019.1646930
7. Lee, W., Jung, H., Bok, I., & Kim, J. (2016). Measurement and application of 3D ear images for earphone design. In *Proceedings of the human factors and ergonomics society annual meeting* (Vol. 60, pp. 1053–1057). SAGE. https://doi.org/10.1177/1541931213601244
8. Binderup, A. T., Arendt-Nielsen, L., & Madeleine, P. (2010). Pressure pain sensitivity maps of the neck-shoulder and the low back regions in men and women. *BMC Musculoskeletal Disorders, 11*, 11–16. https://doi.org/10.1186/1471-2474-11-234
9. Alburquerque-Sendín, F., Madeleine, P., Fernández-De-Las-Peñas, C., et al. (2018). Spotlight on topographical pressure pain sensitivity maps: A review. *Journal of Pain Research, 11*, 215–225. https://doi.org/10.2147/JPR.S135769
10. Jensen, T. S., & Finnerup, N. B. (2014). Allodynia and hyperalgesia in neuropathic pain: Clinical manifestations and mechanisms. *The Lancet Neurology, 13*(9), 924–935. https://doi.org/10.1016/S1474-4422(14)70102-4
11. Fernández-De-Las-Peñas, C., Madeleine, P., Martínez-Perez, A., et al. (2010). Pressure pain sensitivity topographical maps reveal bilateral hyperalgesia of the hands in patients with unilateral carpal tunnel syndrome. *Arthritis Care and Research, 62*(8), 1055–1064. https://doi.org/10.1002/acr.20189
12. Doutres, O., Sgard, F. C., Benacchio, S., et al. (2018). Earplug comfort: From subjective assessment on the field to objective measurement and simulation using augmented artificial heads. *The Journal of the Acoustical Society of America, 143*(3), 1910–1930. https://doi.org/10.1121/1.5036223
13. Guo, Y., Zhang, X. S., Wang, Y., et al. (2018). All-fiber hybrid piezoelectric-enhanced triboelectric nanogenerator for wearable gesture monitoring. *Nano Energy, 48*, 152–160. https://doi.org/10.1016/j.nanoen.2018.03.033
14. Terroir, J., Perrin, N., Wild, P., et al. (2021). Assessing the comfort of earplugs: Development and validation of the French version of the COPROD questionnaire. *Ergonomics, 64*(7), 912–925. https://doi.org/10.1080/00140139.2021.1880027
15. Doutres, O., Sgard, F., Terroir, J., Perrin, N., Jolly, C., Gauvin, C., & Negrini, A. (2020). A critical review of the literature on comfort of hearing protection devices: Analysis of the comfort measurement variability. *International Journal of Occupational Safety and Ergonomics, 28*(1), 447–458. https://doi.org/10.1080/10803548.2020.1772546

16. Wang, M., Fan, H., Yu, S., Zhao, X., Wang, L., Li, W., Wang, L., Yu, M., Chu, J., Zhang, S., & Chen, D. (2022). Effects of variations in the tragus expansion angle on physical comfort for in-ear wearables. *Ergonomics, 65*(10), 1352–1372. https://doi.org/10.1080/00140139.2022.2032377
17. Barnard, R. W., Pearce, K., & Schovanec, L. (2001). Inequalities for the perimeter of an ellipse. *Journal of Mathematical Analysis and Applications, 260*(2), 295–306. https://doi.org/10.1006/jmaa.2000.7128
18. Casali, J. G., & Grenell, J. F. (1990). Noise-attenuating earmuff comfort: A brief review and investigation of band-force, cushion, and wearing-time effects. *Applied Acoustics, 29*(2), 117–138. https://doi.org/10.1016/0003-682X(90)90026-Q
19. Sickel, K., Baloch, S., Melkisetoglu, R., et al. (2011). Toward automation in hearing aid design. *Computer-Aided Design, 43*(12), 1793–1802. https://doi.org/10.1016/j.cad.2011.06.005
20. Fu, F., & Luximon, Y. (2019). Fit and comfort perception on hearing aids: A pilot study. In *International conference on human systems engineering and design: Future trends and applications* (pp. 360–364). Springer. https://doi.org/10.1007/978-3-030-27928-8_55
21. Ball, R., Shu, C., Xi, P., et al. (2010). A comparison between Chinese and Caucasian head shapes. *Applied Ergonomics, 41*(6), 832–839. https://doi.org/10.1016/j.apergo.2010.02.002

Chapter 5
Characterization of Auricle-EAM Transitional Structure and Its Expansion

Abstract Intrusive ear-related wearables rely heavily on the mechanical behavior of the anatomical region connecting the auricle and the external acoustic meatus (EAM), yet this transitional structure has received little systematic investigation. This chapter defines the auricle-EAM transitional structure as an integrated unit consisting of the medial tragus, the anterior cartilage of the medial concha, and the anterior and posterior margins of the EAM entrance, and examines its deformation characteristics and comfort implications under intrusive fitting conditions. A novel angular variable, the angle between tragus, entrance, and medial concha (ATEMC), is proposed to characterize its shape-structure relationship. Using large-scale anthropometric data and controlled expansion experiments, the study demonstrates that ATEMC shows substantial interindividual variability and follows a phasic age-growth pattern, with apparent sex differences largely explained by body-size covariates. To standardize deformation across individuals, the tragus expansion angle (TEA) and the tragus expansion index (TEI) are introduced and linked to subjective comfort ratings. Results reveal a staged comfort-perception process in which functional comfort dominates at low expansion levels, mixed functional-physical perception occurs near the optimal range, and physical discomfort governs responses at higher expansion levels. Quantitative comfort thresholds are identified, providing actionable design criteria for in-ear devices. Overall, the findings establish a robust ergonomic framework that bridges anthropometric measurement and comfort perception, offering practical guidance for the design of stable, comfortable intrusive ear-related wearables.

Keywords Auricle-EAM transitional structure · ATEMC · Tragus expansion index (TEI) · Wearing comfort · Ergonomic design

The ergonomics of ear-related wearable devices has become a critical topic in recent years, particularly as the demand for high-performance in-ear technologies continues to expand across consumer electronics, occupational hearing protection, and biomedical monitoring. While the auricle and the external acoustic meatus (EAM) have been independently examined in terms of anthropometry and ergonomic design, the transitional structure between these regions has not been systematically

H. Fan et al., *Engineering Anthropometry of the External Ear for Human-Product Interaction*, https://doi.org/10.1007/978-981-92-2160-8_5

investigated. This region, defined as the auricle-EAM transitional structure, comprises the medial tragus, the anterior cartilage of the medial conchal midpoint, and the anterior and posterior sides of the EAM entrance. Because intrusive ear-related wearables rely on the stability of this transitional structure for effective fitting, its deformation under loading conditions represents both a determinant of physical comfort and a constraint on product design.

Prior studies have emphasized that physical comfort in ear-related devices is not merely a sensory attribute but an integrative factor that influences user compliance, device effectiveness, and long-term acceptance. For example, research on earplugs has demonstrated that comfort perception arises from a combination of mechanical fit, material interaction, and user-specific anthropometry. Similarly, studies on hearing protectors and ear-related wearables have revealed that small variations in structural dimensions can induce disproportionate changes in perceived discomfort, leading to premature device removal or reduced compliance with protective protocols. The implication of these findings is that comfort is a multidimensional construct that encompasses physical, psychological, physiological, and functional dimensions. In the case of intrusive ear-related wearables, the transitional structure between the auricle and the EAM serves as the locus where these dimensions converge, making its characterization essential for both ergonomics and applied design.

From an anthropometric perspective, auricle and EAM morphology exhibit substantial interindividual variability, which can be attributed to persistent age-growth, phasic age-growth, sex-related differences, and bilateral asymmetry. Investigations of the auricle have indicated that auricular shape categories and dimensions vary systematically with age and sex, thereby influencing the classification of ear types for ergonomic design. Studies of the EAM have similarly reported variations in length, width, and angulation, with implications for device sealing and stability. However, while auricular morphology and EAM dimensions have been studied independently, the structural relationship between the tragus, concha, and meatus entrance (the auricle-EAM transitional structure) has received limited attention. The lack of empirical evidence regarding its deformation and comfort implications constitutes a major gap in the literature, particularly given the widespread application of intrusive devices.

Existing models of comfort in hearing protection devices and wearables have predominantly relied on subjective reports or indirect simulations. For instance, studies employing artificial heads and finite element models have approximated user responses to earplug deformation, yet these approaches cannot capture the nuanced perceptual feedback that arises from transitional structural expansion. Moreover, the measurement variability and definitional ambiguities surrounding comfort have been repeatedly noted in critical reviews, underscoring the need for standardized characterization methods. By contrast, the auricle-EAM transitional structure can be operationalized through measurable angular variables, specifically the angle between tragus, entrance, and medial concha (ATEMC), and its derived indices such as the tragus expansion angle (TEA) and the tragus expansion index (TEI). These variables provide a quantifiable framework for linking structural deformation to perceptual outcomes.

Research is needed to bridge the gap between anthropometric measurements and comfort perception. On one hand, product designers require reliable metrics that predict user responses under intrusive conditions. On the other hand, ergonomic science demands robust evidence that delineates the exposure-response relationship between transitional structure expansion and comfort perception. The integration of both objectives is particularly relevant in the design of earphones, hearing aids, communication headsets, and clinical diagnostic tools, where intrusive fitting is unavoidable. Without accurate characterization of the auricle-EAM transitional structure, device design remains susceptible to trial-and-error approaches, which compromise both efficiency and user welfare.

5.1 Measurement and Characterization of the Auricle-EAM Transitional Structure

5.1.1 Expansion Deformation of the Auricle-EAM Transitional Structure

The descriptive variables corresponding to the auricle-EAM transitional structure involve the anatomical characteristics of the tragus and the entrance of the external acoustic meatus. In the semi-structured interview, participants report that when they use intrusive ear-related wearables such as plug-type devices, band-type devices, or in-ear components, the tragus undergoes outward expansion that is associated with product fitting (Fig. 5.1). When the tragus expands outward, the entrance of the external acoustic meatus and the concha frequently produce varying degrees of discomfort, particularly when products with larger dimensions are applied. In the structured questionnaire survey, the dimensional variables of the concha or the cavum concha, which are also defined as descriptive variables, are not categorized

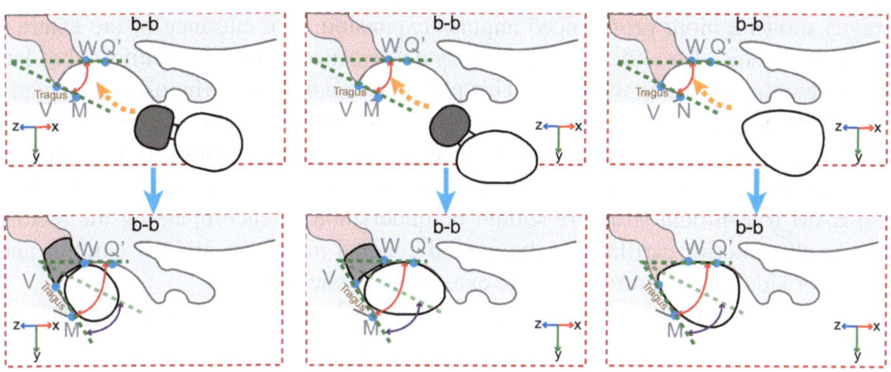

Fig. 5.1 Tragus expansion angle: tragion (M), medial concha (Q), anterior entrance (V), posterior entrance (W), medial concha (projection) (Q')

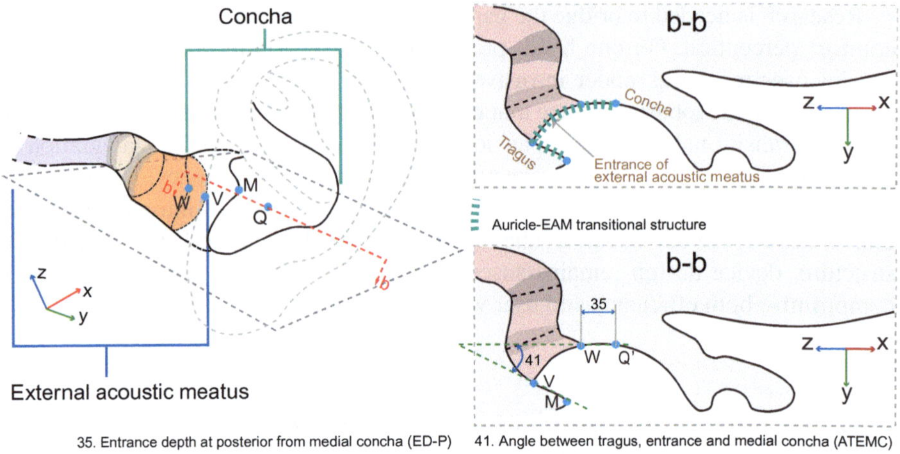

Fig. 5.2 Variable of characterization for auricle-EAM transitional structure: angle between tragus, entrance and medial concha (41); auxiliary variable and landmarks include entrance depth at posterior from medial concha (35), tragion (M), medial concha (Q), anterior entrance (V), posterior entrance (W), medial concha (projection) (Q')

into the auricle-EAM transitional structural factor, which can be plausibly explained by the fact that in actual discomfort experiences, participants' perception of conchal discomfort is neither immediate nor pronounced.

The anterior and posterior sides of the entrance of the external acoustic meatus are respectively connected with the medial surface of the tragus and the anterior cartilage of the medial midpoint of the concha. The auricle-EAM transitional structure specifically refers to this connecting region (Fig. 5.2), and the structural landmarks, reference axes, and planes remain consistent with those described in Chap. 3 [1]. Under this anatomical arrangement, when intrusive devices push the medial tragus and the concha anteriorly and posteriorly, a relatively large angular relationship is formed. Because the anterior cartilage of the medial conchal midpoint is closely attached to the skull, it exhibits little or no outward expansion, whereas the tragus shows a more pronounced angular expansion. The entrance of the external acoustic meatus also undergoes temporary deformation, which is manifested as lateral stretching along its width [2]. This phenomenon explains why users frequently adjust the position of intrusive devices when discomfort occurs, since such adjustments represent attempts to restore balance in their perceived comfort. Accordingly, in terms of both structural characteristics and product fitting mechanisms, the auricle-EAM transitional structure should be understood as encompassing the medial tragus, the anterior cartilage of the medial conchal midpoint, and the anterior and posterior sides of the entrance of the external acoustic meatus.

The auricle-EAM transitional structural factor, which is categorized as Ergonomic Design Factor 3, requires investigation through two measurement variables: the angular variable of the transitional structure itself and the expansion angle of the tragus. When the auricle-EAM transitional structure expands, the exposure of skin surface-pressure sensitivity at the medial tragus is induced by the lateral mechanical pressure exerted by intrusive devices (Fig. 5.1). The negative mechanical effects of intrusion are therefore concentrated primarily on the tragus, which results in a significant correlation between the two variables [3, 4]. Research on the exposure-response relationship of auricle-EAM transitional structure expansion accordingly focuses on the angular variable of the transitional structure itself and on the tragal expansion angle generated by the fitting of in-ear components.

5.1.2 Shape-Structure Characterization and Measurement Variable

For the auricle-EAM transitional structure, which is categorized as Ergonomic Design Factor 3, the shape-structure characterization requires one measurement variable. This variable is defined as the angle between tragus, entrance and medial concha (ATEMC), which represents the angular relationship among the medial tragus, the anterior cartilage of the medial conchal midpoint, and the anterior and posterior points of the entrance of the external acoustic meatus (Fig. 5.2). The outcome of this measurement variable is applicable to the design of the connection between in-ear components and conchal components, and it is also relevant to the overall cross-sectional morphology of conchal components [5]. The correspondence between the descriptive variables and the measurement variables is summarized in Table 5.1. The auricle-EAM transitional structure is characterized by the measurement variable ATEMC, which is defined as the angle between tragus, entrance and medial concha. Descriptive statistical analysis, correlation analysis, and difference analysis together provide the foundation for defining the tragal expansion angle, for determining wearing comfort thresholds, and for constructing exposure-response models.

5.1.3 Characterization Methods

The characterization study of the auricle-EAM transitional structure is conducted in parallel with the shape-structure methods described in Chap. 3. Participant recruitment, landmark annotation, orientation and reference axis definitions, reference plane construction, and data collection and processing procedures all follow the same principles that are employed in Chap. 3.

Participants are recruited from eight sampling cities, yielding a total of 1120 individuals. Recruitment criteria, participant composition by sex and age group, as well as the distribution of recruitment sites and participants' native provinces are described in Chap. 3. As illustrated in Fig. 5.2, the landmarks employed include the tragion (M), the medial conchal midpoint (Q), the anterior entrance of the external

Table 5.1 Characterization variable for the auricle-EAM transitional structure

No.	Variable	Abbr.	Type
41	Angle between tragus, entrance and medial concha	ATEMC	Angular

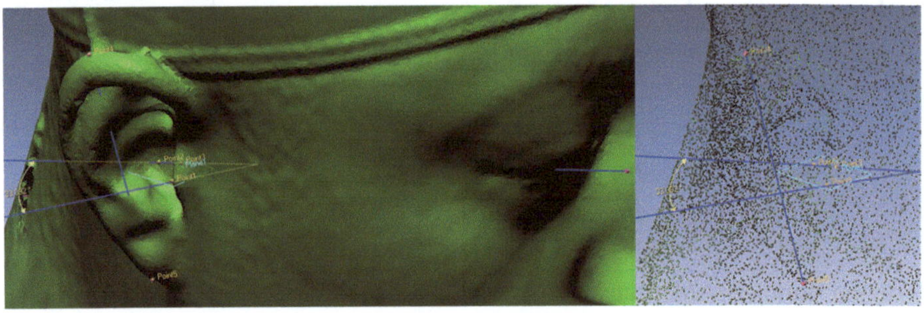

Fig. 5.3 Relative position of ear horizontal plane on the head-face model

acoustic meatus (V), and the posterior entrance of the external acoustic meatus (W). The definitions and creation methods of these landmarks are consistent with those that are described in Chap. 3. The orientation employed is the auricular inclination angle (AIA) orientation. The reference axes include the auricular long axis and the transverse axis of the auricle at the tragion. The reference planes comprise the ear horizontal plane (Fig. 5.3), the Frankfurt horizontal plane, the coronal plane, and the sagittal plane, and the details regarding orientation, reference axes, and reference planes are provided in Chap. 3.

The measurement variable ATEMC is defined as follows: ATEMC (variable 41) is the angle between ED-P (Variable 35) and the medial surface of the tragus, represented by line segment MV, which is parallel to the ear horizontal plane. The procedures for three-dimensional data acquisition of the head-face region and both ears, landmark annotation, and intra-measure reliability evaluation are consistent with those described in Chap. 3. The measurement of the angular variable is conducted using Eq. (3.2), which calculates the angle between line segment Q′W and line segment MV (Figs. 5.2 and 5.3), where Q′ represents the vertical projection of the medial conchal midpoint (Q) onto the ear horizontal plane.

Statistical analyses are performed in MATLAB under a 95% confidence level ($\alpha = 0.05$), and the results are regarded as statistically significant when $p < 0.05$. One-way ANOVA is employed to assess the overall influence of age groups on the characterization variable, and Tukey's HSD test is used for post hoc multiple comparisons across age groups. The Pearson correlation coefficient is applied to examine correlations between the characterization variable and participants' age, height, weight, and BMI. Paired-sample t-tests and independent-sample t-tests are adopted to analyse bilateral differences and sex-related differences of the characterization variable, and when independent-sample t-tests reveal significant differences, height and weight serve as covariates to adjust the corresponding differences, while Bonferroni correction is applied for further post hoc comparisons. Based on the outcomes of age-group difference analysis and growth-trajectory analysis, linear or nonlinear fitting methods are applied to establish either a phasic age-growth model or a persistent age-growth model of ATEMC.

5.1.4 *Ethics Approval*

The studies of this chapter performed in line with the principles of the Declaration of Helsinki. Informed consent was obtained from all participants. The studies of this chapter were approved by the Ethics Committee of Northwestern Polytechnic University (No. IRB No. 201502024, No. 202002024).

5.2 Characterization Outcomes

5.2.1 *Descriptive Statistics and Correlation Analysis*

As summarized in Table 5.2, the range of ATEMC is 16.38° to 58.70°, with a mean value of 31.66°. Significant differences are identified across age-group distributions ($p < 0.001$), and the percentile statistics are presented in Table D.1. The correlations between the measurement variable and age, height, weight, and BMI are summarized in Table 5.3, and the measurement variable ATEMC shows no significant correlation with height. This result indicates that participants of older age or higher body mass tend to exhibit a greater baseline expansion of the auricle-EAM transitional structure.

5.2.2 *Phasic Age-Growth Model*

Table 5.4 reports the multiple comparison results of ATEMC based on age-group distributions, and the detailed numerical outcomes are presented in Table D.2. The results indicate that this variable belongs to the category of phasic age-growth, and non-significant increases are still observable after 69 years of age ($p > 0.05$).

Table 5.2 Descriptive statistics of measurement variable for the auricle-EAM transitional structure

No.	Variable	Ear, N = 2240		Max	Min	ANOVA	
		Mean	SD			df (age group)	p
41	ATEMC (°)	31.66	6.42	16.38	58.70	6	< 0.001

Table 5.3 Correlation analysis of measurement variable for the auricle-EAM transitional structure

		Age		Height		Weight		BMI	
No.	Variable	Pearson correlation	Sig. (2-tailed)	Pearson correlation	Sig. (2-tailed)	Pearson correlation	Sig. (2-tailed)	Pearson correlation	Sig. (2-tailed)
41	ATEMC	0.43[a]	< 0.001	−0.04	0.19	0.11[a]	< 0.001	0.15[a]	<0.001

[a] Correlation is significant at the 0.01 level (2-tailed)

Table 5.4 Multiple comparisons of measurement variable for the auricle-EAM transitional structure (by age groups)

		Multiple comparison across age group			
No.	Variable	a	b	c	d
41	ATEMC	10s, 20s	30s	40s, 50s	60s, 70s

a, b, c, d: multiple comparison results of Tukey test, the same superscript on mean values indicated no significant mean differences between age groups. α = 0.05

Table 5.5 Comparison of difference of male-female and left-right in measurement variable for the auricle-EAM transitional structure

		Male-female (independent samples)					Left-right (paired samples)			
No.	Variable	Gender	Mean	SD	Mean different	Adjusted Mean different[a]	Side	Mean	SD	Mean different
41	ATMEC (°)	Male	32.10	6.71	0.88[b]	0.55	左	31.72	6.39	0.20
		Female	31.22	6.13			右	31.52	6.45	

[a] Covariates are evaluated at the following values: Stature (mean) = 1.67, Weight (mean) = 63.90. Adjustment for multiple comparisons: Bonferroni
[b] $p < 0.05$, α = 0.05, significant values are marked with the grey colour

5.2.3 *Bilateral and Sexual Differences*

As shown in Table 5.5, ATEMC exhibits significant sex-related differences, whereas the bilateral differences are not significant. After the effects of height and weight are adjusted, the sex-related differences of ATEMC are no longer significant. In subsequent analyses, the mean value of ATEMC continues to be employed for the establishment of the wearing comfort thresholds, the exposure limits, and the related quantitative models.

5.2.4 *Growth Trajectory*

Because the auricle-EAM transitional structure functions as the connecting region between the auricle and the external acoustic meatus, it is not included in the scope of the analyses described in Chap. 3. When ATEMC is regarded as a factor associated with shape-structure variation, the growth trajectory of the auricle-EAM transitional structure can be described as follows: the value of the angle between tragus, entrance and medial concha (ATEMC) increases persistently with age. This indicates that, along with the continuous enlargement of the entrance width of the external acoustic meatus, the tragus gradually deviates laterally toward the concha and inclines toward the coronal plane of the head-face system.

5.2.5 Phasic Age-Growth Model

ATEMC belongs to the category of phasic age-growth measurement variables, and its curve relationship with age is characterized by a monotonically increasing trajectory. According to the methodological framework described in Chap. 4, the relationship between ATEMC and age is fitted using age means. Among the candidate models, the Logistic E-R function (Eq. (4.2)) achieves the best fitting performance (adjusted $R^2 = 0.802$). The fitting outcome is presented in Eq. (5.1), and the curve-fitting result is illustrated in Fig. 5.4.

$$ATEMC = 35.66659 - \frac{8.82158}{1 + \left(\frac{Age}{38.4615}\right)^{3.1007}} \tag{5.1}$$

where, $Age \in [15, 79]$.

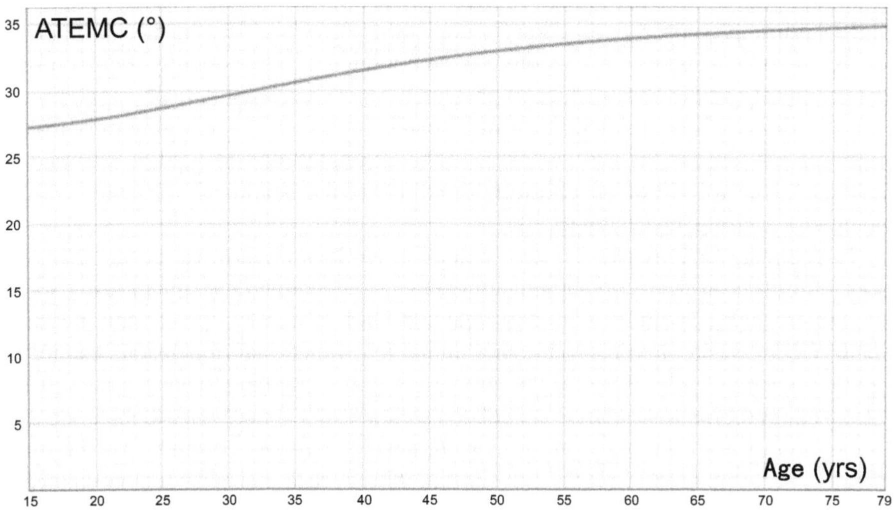

Fig. 5.4 Age-variable curve fitting (ATEMC)

5.3 Comfort Perception in Intrusive Ear-Related Wearables

5.3.1 Functional Dimension of Global Comfort Perception

When referring to human comfort in the context of ear-related wearables, unless otherwise specified, the term denotes the wearing comfort of the external ear. According to the Naddeo-Capetti conceptual equation (Eqs. (1.1) and (1.2)), users' global comfort perception of wearables includes four basic dimensions: the physical dimension, which concerns mechanical pressure on the skin surface, body dimensions, joint angles, and muscular activity; the psychological or cognitive dimension, which concerns visual appearance, aesthetic experience, psychological stress, and satisfaction; the physiological dimension, which concerns physiological changes after use, such as heart rate and blood pressure; and the environmental or macro-ergonomic dimension, which concerns sound, thermal conditions, light, humidity, vibration, and organizational or systemic density or activity. User perception of these four dimensions is reflected in ear-related wearables as perception of wearing comfort, psychological or cognitive comfort, physiological comfort, and environmental or macro-ergonomic comfort.

Wearables are often expected to provide users with enhanced or augmented performance. Accordingly, in addition to the basic dimensions, perception of the functional dimension should also be considered, depending on the specific design form of the wearable, which determines whether the functional qualities or effects provided by the device restrict normal user activity. The functional dimension of ear-related wearables is described in terms of device weight, sound quality or noise-reduction level, and the mode of contact with the external ear, all of which may constrain normal activities.

In studies of comfort perception for non-intrusive wearables, the measurement of the basic and functional dimensions can be conducted independently. For example, the sensitivity of the skin surface to pressure, which represents the physical dimension, in users of noise-reduction earmuffs can be measured by applying a controlled compressive force to the skin, whereas the functional dimension can be assessed by increasing the weight of the earmuffs or adjusting the headband height to evaluate whether activity is restricted [6, 7].

In contrast, for intrusive devices such as in-ear headphones, ear-canal EEG sensors, or oral-nasal contact breathing apparatus for underwater use, the intrusive mode of contact itself constitutes a fundamental condition of comfort-perception measurement. Under these circumstances, the human and functional dimensions cannot be studied independently. Because the auricle-EAM transitional structure serves as the fitting target region for intrusive ear-related wearables, it is necessary to integrate user perception of both the physical dimension, namely wearing comfort, and the functional dimension, namely functional comfort, which are associated with the expansion of this structure. On the basis of these results, the expansion-perception process can be divided into distinct stages that are controlled by the relevant dimensions.

5.3.2 Definition of Comfort Perception in Intrusive Ear-Related Wearables

The global comfort perception of intrusive ear-related wearables is governed by a multidimensional structure that comprises five dimensions, which include the four basic dimensions, namely the physical, psychological or cognitive, physiological, and environmental or macro-ergonomic dimensions, as well as the functional dimension.

The main attributes of the physical dimension in intrusive ear-related wearables are pain sensation, static mechanical pressure resulting from fitting, and skin irritation caused by differences in design materials or design dimensions. Accordingly, the perception of wearing comfort, or the perception of the physical dimension, is specifically defined as users' perception of pain or irritation that is caused by external stimuli due to differences in shape, material, or design dimensions of the measuring tool, or due to static mechanical pressure resulting from its fitting on the external ear.

Since the measuring instruments employed in this study are also intrusive devices, smaller instruments may approximate more closely the dimensions of the user's own auricle-EAM transitional structure, whereas larger instruments may impose weight that restricts head movement. Therefore, the perception of functional comfort, or the perception of the functional dimension, is specifically defined as users' perception of the fitting stability of the measuring tool, which concerns whether the position can be maintained, and its intrusiveness, which concerns whether head movement is restricted, during the fitting on the external ear [8–15]. On the basis of this definition, the results that are influenced or controlled by the functional dimension are identified and distinguished.

5.3.3 Tragus Expansion Angle

When intrusive devices are fitted to the auricle-EAM transitional structure, the resulting expansion is primarily manifested as a pronounced outward displacement of the tragus. The angular value of the measurement variable ATEMC increases in proportion to the degree of tragal expansion, thereby creating a difference from the user's original ATEMC angle (Fig. 5.5). This angular difference is defined as the tragus expansion angle (TEA). TEA is specifically expressed as the angular difference between the post-fitting ATEMC value generated by the measuring tool ($ATEMC_{wear}$) and the user's original ATEMC value ($ATEMC_{ear}$) (Eq. (5.2)).

$$TEA = \Delta ATEMC = ATEMC_{wear} - ATEMC_{ear} \tag{5.2}$$

where, TEA is the Tragus Expansion Angle, $ATEMC_{wear}$ is the post-fitting ATEMC value generated by the measuring tool; $ATEMC_{ear}$ is the user's original ATEMC value; $ATEMC_{wear} \geq ATEMC_{ear}$, $\Delta ATEMC \geq 0$.

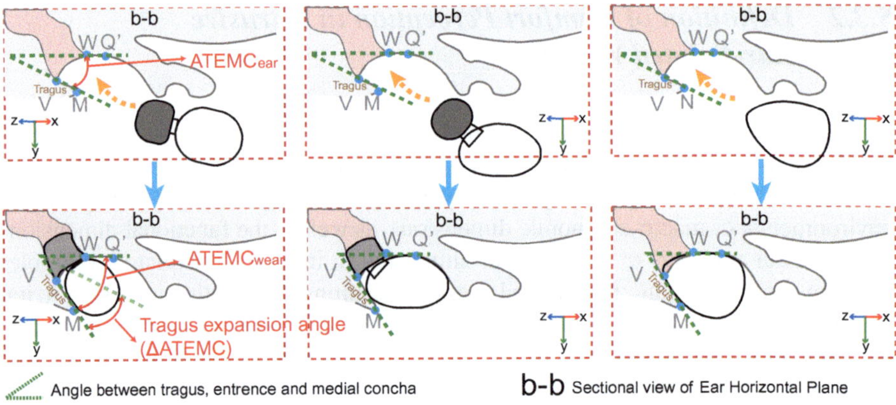

Fig. 5.5 Tragus expansion angle (ΔATEMC) and angle between tragus, entrance and medial concha: tragion (M), medial concha (Q), anterior entrance (V), posterior entrance (W), medial concha (projection) (Q')

5.3.4 Tragus Expansion Index

Because the values of ATEMC vary among participants (Table 5.1), an identical TEA value, for example 1°, does not correspond to the same degree of structural expansion across individuals. To standardize and to describe the extent of expansion corresponding to the measured TEA values, the tragus expansion index (TEI) is introduced. TEI is specifically defined as the ratio of a participant's measured TEA value to the individual's own ATEMC value, multiplied by 100% (Eq. (5.3)).

$$TEI = \frac{TEA}{ATEMC_{ear}} \times 100\% = \frac{\Delta ATEMC}{ATEMC_{ear}} \times 100\% \tag{5.3}$$

where, *TEA* is the Tragus Expansion Angle, $TEI \geq 0$; $ATEMC_{ear}$ is the user's original ATEMC value; $\Delta ATEMC \geq 0$.

5.3.5 Design of the ATEMC Measurement Tool

Design

The measurement objective is to capture the tragus expansion angle (TEA) that is generated when intrusive ear-related wearables are fitted to the external ear. Therefore, the cross-sectional geometry of the speaker rubber rings, or the soft tips of in-ear components is not regarded as a design consideration for the measurement tool (Fig. 5.6). The ATEMC measurement tool that is designed and employed for this purpose is illustrated in Fig. 5.7, and its design objective is to simulate the

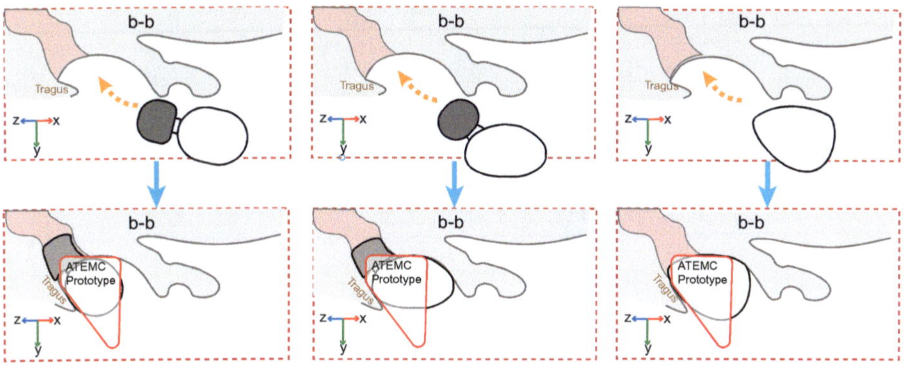

Fig. 5.6 Fitting relationships of the cross-sectional shapes between ATEMC prototypes and in-ear wearables

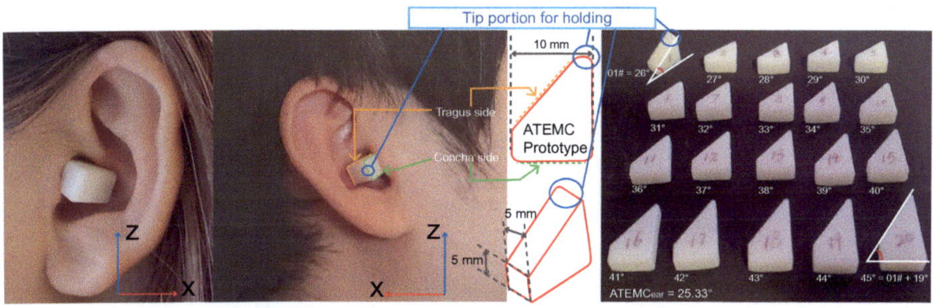

Fig. 5.7 ATEMC prototypes

fitting process of intrusive ear-related wearables at the auricle-EAM transitional structure and to measure TEA directly. All ATEMC measurement tools are fabricated using stereolithography apparatus (SLA) printing with the resin material DSM SOMOS® Imagine-8000 (accuracy: 0.2 mm; hardness/Shore D: 79; density: 1.16 g/cm^3). Because of its effectively low weight load, the Imagine-8000 material is widely applied in clinical medicine, surgical planning, aerospace engineering, and consumer product research and design, and its use here aims to minimize the potential influence of material weight on participants' functional comfort perception during experiments.

The ATEMC measurement tools are manufactured at different angles to simulate varying degrees of expansion of the auricle-EAM transitional structure. Depending on the specific angle, the final products weigh between 1.7 g and 3.0 g. The angle formed between the concha side and the tragus side of the ATEMC measurement tool is defined as $ATEMC_{wear}$ (Fig. 5.7). To enhance the clarity, accuracy, and controllability of TEA measurement, the thickness of the tool is designed to be 5 mm. Additional details are provided in the subsequent section on the historical version of the design scheme. The ATEMC measurement tool is designed with rounded

Fig. 5.8 ATEMC prototype (rounded corners)

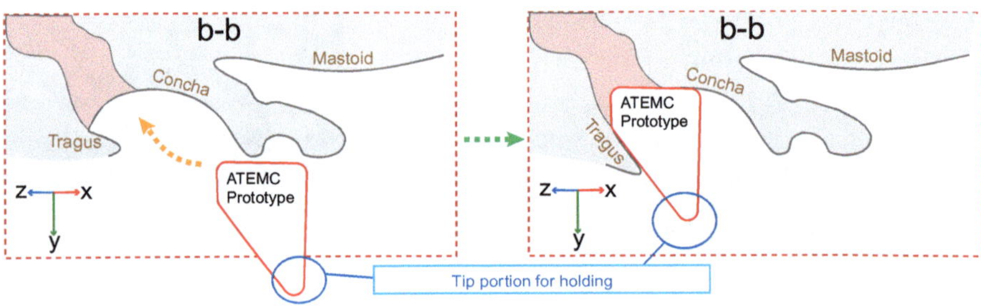

Fig. 5.9 Wearing process of ATEMC prototype

corners (R = 0.5 mm, Fig. 5.8) and a tapered tip section for manual adjustment of the fitting position (Fig. 5.9). The top and bottom surfaces of the tool are congruent rounded-corner right trapezoids, each with an upper base length of 5 mm.

The auxiliary variable ED-P, which is defined as the depth from the posterior entrance point of the external acoustic meatus to the medial conchal midpoint, is employed to determine the length of the concha side, namely the right-angle edge on the top and bottom surfaces, of the ATEMC measurement tool. Since the ED-P values of the participants recruited for the structural-expansion study are all greater than 10 mm, the concha-side design value of the ATEMC measurement tool is uniformly set at 10 mm in order to ensure full contact with the auricle-EAM transitional structure.

For each invited participant, 20 ATEMC measurement tools with different angles for a single external ear are designed and fabricated, along with 10 additional tools of larger angles as provisional designs. Tools 1# (weight $\approx$ 1.7 g) through 20# (weight $\approx$ 3.0 g) are employed for actual measurements, whereas tools 21# (weight $\approx$ 3.0 g) through 30# (weight $\approx$ 3.4 g) serve as backup schemes. When the first 20 tools do not elicit extreme discomfort perception in a participant, and with participant consent, the additional 10 tools are fabricated and used in a follow-up measurement session.

For each participant's unilateral external ear, beginning with tool 2#, the $ATEMC_{wear}$ value of each subsequent tool increases by 1° relative to that of the preceding tool. The $ATEMC_{wear}$ value of tool 1# for each ear is determined by the participant's measured ATEMC_ear value. Due to limitations of 3D-printing precision, the $ATEMC_{wear}$ value of tool 1# cannot be made exactly identical to the measured $ATEMC_{ear}$ value. To address this limitation, the design value of tool 1# is rounded up to the next integer above the measured $ATEMC_{ear}$ value. For example, if the measured $ATEMC_{ear}$ value of a participant's left ear is 25.33°, the design value of tool 1# is 26°, tool 20# is 45°, and the provisional tool 30# is 55°.

Individual differences in measurement variables of the external ear are evident. Some participants exhibit insensitive perceptual responses to the expansion of the auricle-EAM transitional structure within a certain range, whereas others are able to tolerate devices with values considerably greater than their own baseline. On the basis of these considerations, 10 provisional tools (21#–30#) are prepared for each invited participant as a backup scheme. An overview of the dataset composition is presented in Fig. 5.10.

Testing and Versions

The recommended design scheme and version of the ATEMC measurement tool is the T5 version (5-mm thickness) of the rounded-rectangle scheme. In this version, the cross-sections that face the external acoustic meatus and the opposite side of the external ear are rectangular, and the section that faces the external acoustic meatus is designed as a rounded square of 5 mm × 5 mm (Fig. 5.8). As shown in Fig. 5.11, the ATEMC measurement tool is developed in two schemes with a total of eight versions. In addition to the rounded-rectangle scheme, an elliptical scheme is tested, in which the cross-sections that face the external acoustic meatus and the opposite side of the external ear are elliptical, and the section that faces the meatus is designed as a circle with a 5-mm diameter.

Both schemes include thickness versions T4 (4 mm), T5 (5 mm), T6 (6 mm), and T7 (7 mm). Before the start of formal measurement, these eight versions are subjected to laboratory testing. A total of 15 participants (meeting the required laboratory sample size; 7 males and 8 females, aged 21–26 years) are recruited. For each participant, measurement tools 1#–20# are designed on the basis of the participant's bilateral ATEMC values. During the testing of the T6 and T7 versions in both schemes, three and four female participants, respectively, report that the tools cannot be used even at tool 1# because the thickness exceeds the length or width of the external acoustic meatus. When testing the T4 versions of both schemes, the tragus side or the concha side of tool 1# causes slight discomfort upon contact with the auricular soft tissues; this discomfort is not attributable to tragal expansion and is reported by 11 participants. When testing the T5 version of the elliptical scheme, the

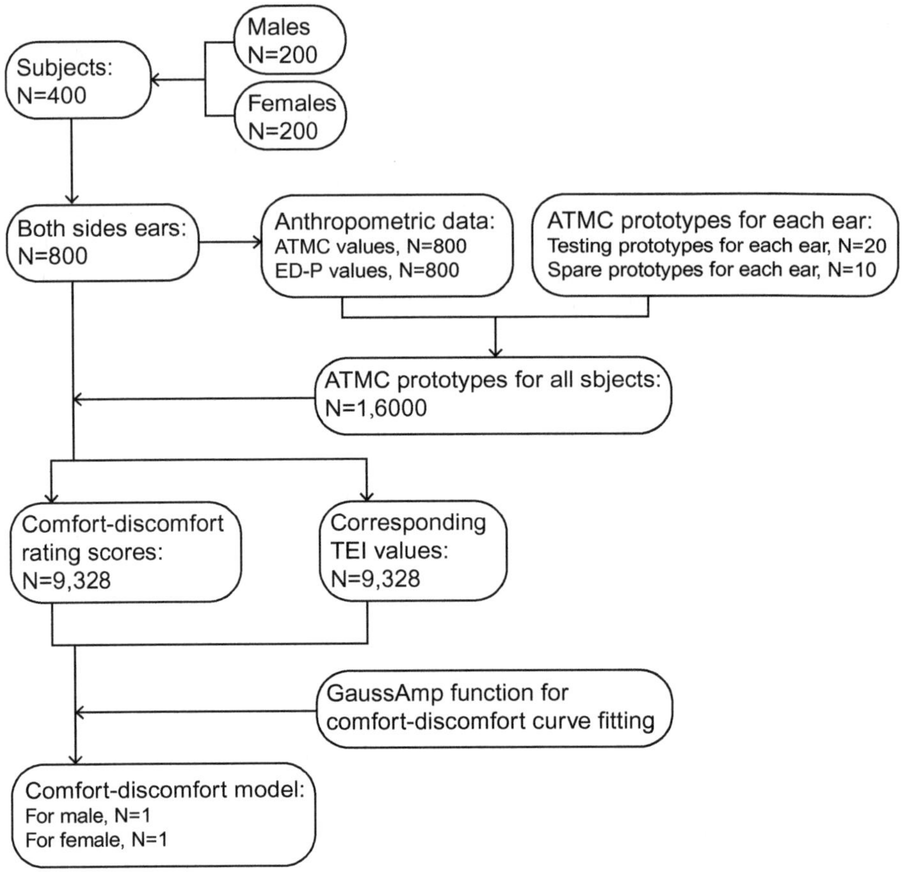

Fig. 5.10 Descriptions of subjects and data (ATEMC: angle between tragus and medial concha; ED-P: entrance depth at posterior from medial concha; GaussAmp: amplitude version of Gaussian peak function)

concha side of tool 1# directly causes slight discomfort in the concha, again unrelated to tragal expansion. On the basis of these laboratory testing results, the T5 version of the rounded-rectangle scheme is selected as the final version for the formal study. In the testing study, the specific auricular regions where tool 1# of each scheme and version elicit discomfort are illustrated in Fig. 5.11.

5.3.6 Design of the Tragus Expansion Comfort Scale

Comfort Levels

In non-comparative studies, the baseline physical comfort value prior to product interaction is generally defined as 0, which means that before the occurrence of instrumental activity the user experiences no discomfort. Once interaction with a product occurs,

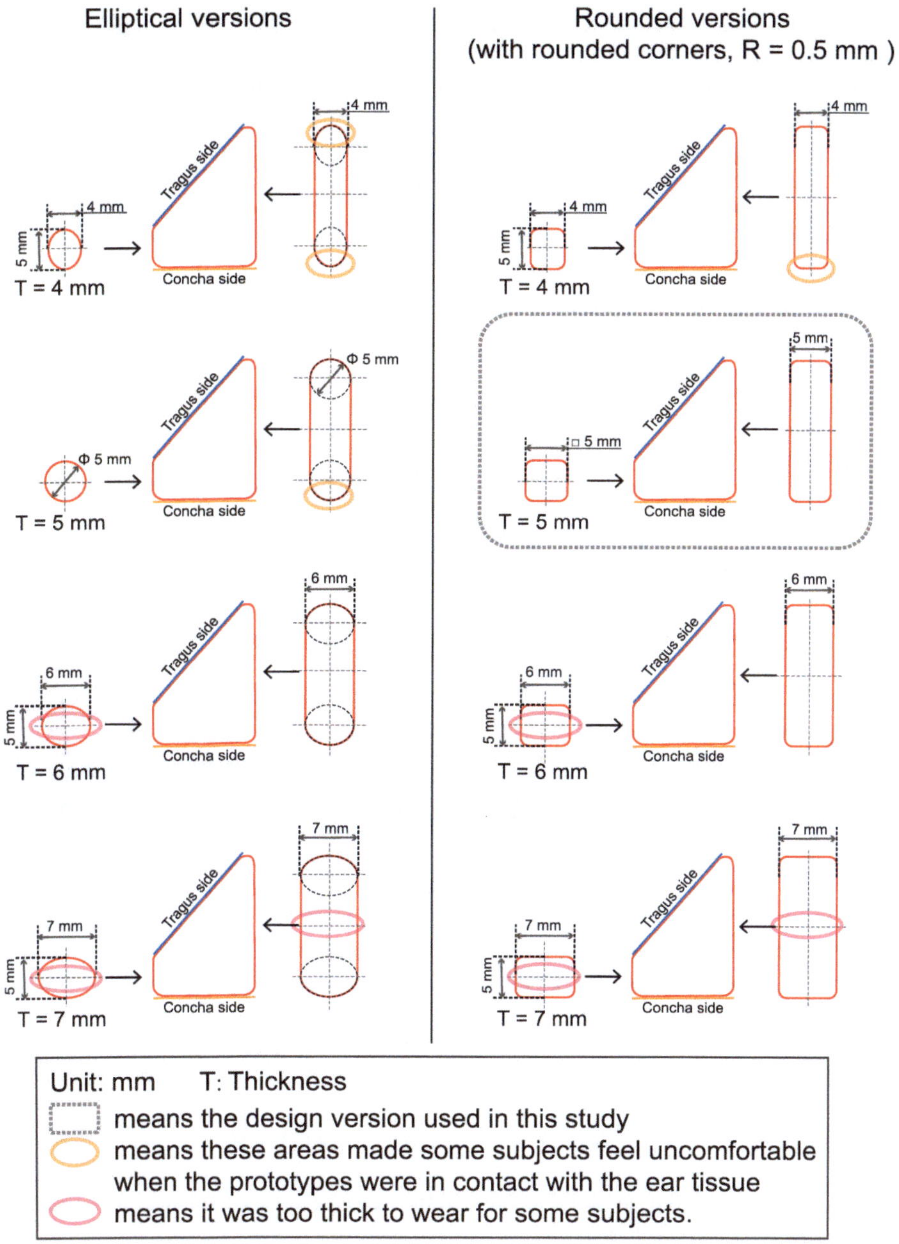

Fig. 5.11 Design versions of ATEMC prototype

negative feedback in physical comfort perception arises, and comfort begins to diminish [16]. In the same context, functional comfort perception is shaped by users' habitual experiences and tends to involve passive comparison. Accordingly, in an idealized state, users' functional comfort perception provides fully positive feedback, and this state is typically assigned a value of 5, which represents extreme comfort.

By reference to the 5-point Likert scale [17] and the Borg CR-10 scale [18], a mixed Tragus Expansion Scale (Fig. 5.12) is designed to measure participants' combined perceptions of both the physical and functional dimensions during tragus expansion. The design scheme is as follows:

Subject:________ Gender:________ Side of ear: ________

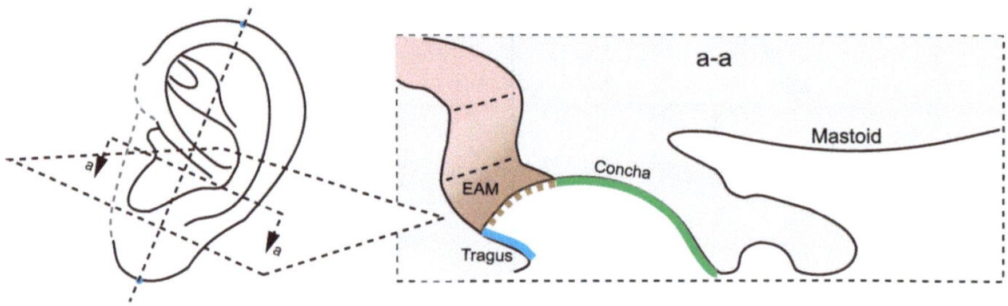

	Just noticeable discomfort	Little discomfort		Moderate discomfort		High discomfort		Very high discomfort		Extreme discomfort
0	-0.5	-1	-1.5	-2	-2.5	-3	-3.5	-4	-4.5	-5
Normal	0.5	1	1.5	2	2.5	3	3.5	4	4.5	5
	Just noticeable comfort	Little comfort		Moderate comfort		High comfort		Very high comfort		Extreme comfort

1. Please record the scale of each No. of ATMC prototype, and stop recording after the one caused ***extreme discomfort*** of **Tragus** (***blue area***). If none of first 20 caused extreme discomfort, move on to 21-30.
01# ( ), 02# ( ), 03# ( ), 04# ( ), 05# ( ), 06# ( ), 07# ( ), 08# ( ), 09# ( ), 10# ( ),
11# ( ), 12# ( ), 13# ( ), 14# ( ), 15# ( ), 16# ( ), 17# ( ), 18# ( ), 19# ( ), 20# ( ).
Supplement:
21# ( ), 22# ( ), 23# ( ), 24# ( ), 25# ( ), 26# ( ), 27# ( ), 28# ( ), 29# ( ), 30# ( ).

2. Please record the first No. and its scale of ATMC prototype which caused the ***discomfort*** of **Concha** (***green area***): ( )#, ( ).

3. Please record the first No. and its scale of ATMC prototype which caused the ***discomfort*** of **External Acoustic Meatus** (***brown area***): ( )#, ( ).

Fig. 5.12 Tragus-expansion rating scale

1. To capture greater sensitivity in participant perception, the 5-level design of the 5-point Likert scale is applied. Five categories are used to measure overall changes in perception across both the physical and functional dimensions throughout the process: "little (±1)," "moderate (±2)," "high (±3)," "very high (±4)," and "extreme (±5)."
2. To accommodate the angular increments of 20 measurement tools for each ear, further optimization is conducted with reference to the Borg CR-10 scale. Each comfort or discomfort level is expanded with an additional intermediate level of 0.5, and two supplementary categories are introduced: "normal (0)" and "ambiguous (±0.5)."

On this basis, the final structure of the comfort-value levels for the Tragus Expansion Scale comprises the following: extreme comfort (5), very high comfort (4.5 and 4), high comfort (3.5 and 3), moderate comfort (2.5 and 2), little comfort (1.5 and 1), just noticeable comfort (0.5), normal (0), just noticeable discomfort (−0.5), little discomfort (−1 and −1.5), moderate discomfort (−2 and −2.5), high discomfort (−3 and −3.5), very high discomfort (−4 and −4.5), extreme discomfort (−5).

Duration of a Single Measurement

The physical comfort level is a critical consideration in the design of tools, particularly those that require continuous contact with the human body [19]. Within a limited period of interaction, users' comfort experience is directly influenced by human-product interaction, namely the differences between the product's design parameters and the user's ergonomic measurements [20]. Accordingly, in the structural-expansion study, the measurement duration for each numbered ATEMC measurement tool depends on the time required for participants to make an initial judgment of whether the corresponding TEA angle is comfortable or uncomfortable.

A survey of all invited participants is conducted on the duration of a single measurement. Approximately 82% of participants report that they are able to make an initial decision within 3–10 min, whereas the remaining participants require 10–15 min. Therefore, the upper limit for a single measurement is set at 15 min, and the duration range for each participant is defined as 3–15 min. The final measurement results are also used to verify whether participants can distinguish the specific comfort values of multiple ATEMC measurement tools within a maximum of 15 min per trial.

5.4 Methods of Expansion Measurement

The objective of the structural-expansion study is to establish a quantitative relationship between the expansion of the auricle-EAM transitional structure and physical comfort perception by standardizing the degree of structural expansion. The specific goals are as follows: to design and to employ the ATEMC measurement tool and the tragus expansion index (TEI) in order to measure and to standardize angular changes of TEA; to design and to employ the Tragus Expansion Scale in order to measure changes in wearing comfort perception and functional comfort perception that are induced by TEA; and to segment the overall expansion-perception process into dimension-control stages on the basis of the measurement results.

5.4.1 Participants and Measurement of Variables

A total of 400 participants is recruited for the structural-expansion study. Participants are aged between 15 and 59 years and are grouped into five age categories: teens (15–19), twenties (20–29), thirties (30–39), forties (40–49), and fifties (50–59). Each age group includes 100 participants with an equal male-to-female ratio. Fifty participants (male-to-female ratio 1:1) are recruited at each of the eight sampling sites: Shenyang, Tianjin, Xuzhou, Shanghai, Fuzhou, Guangzhou, Xi'an, and Lanzhou. The distribution of participant recruitment locations and native provinces is illustrated in Fig. 5.13.

All recruited participants have external ears that are free from through-type piercings, congenital malformations, trauma, corrective surgery, or maxillofacial abnormalities. Familial and kinship restrictions are consistent with the requirements specified in Table 2.2. In addition, to avoid hypersensitivity to pain or cumulative

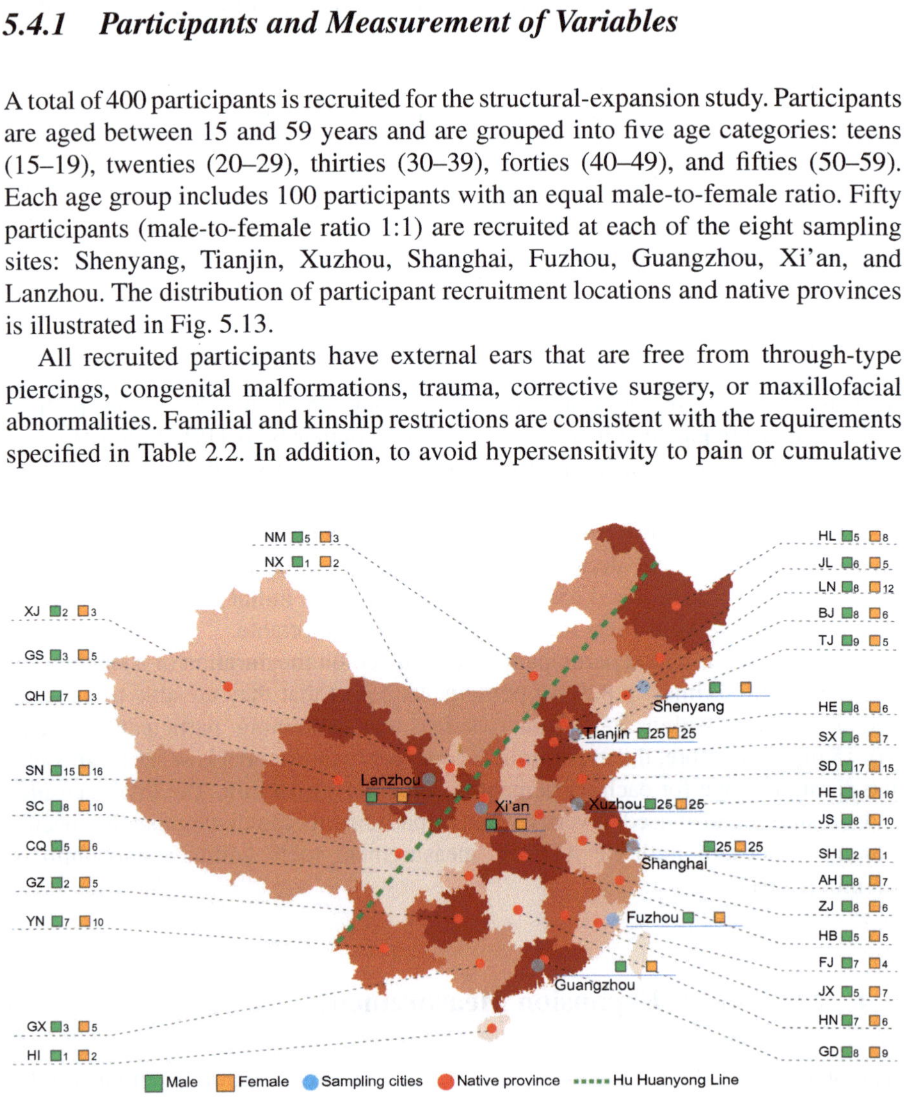

Fig. 5.13 Distribution of participants across sampling locations and native provinces in the structural-expansion study

mechanical pain symptoms during experiments, participants who regularly or continuously take pain-suppressing medications during the previous 12 months, or who experience head-facial pain within the same period, are excluded.

The data collection cycle lasts 12 months. Bilateral auricles of all participants are molded using AB impression materials and three-dimensionally scanned. Procedures for three-dimensional data collection, landmark annotation, and intra-measure reliability evaluation are consistent with those described in Chap. 3. The three-dimensional coordinates of the corresponding landmarks are exported into MATLAB, and ATEMC (design variable) and ED-P (auxiliary variable) are measured using Eqs. (3.1) and (3.2), respectively.

5.4.2 Expansion Data Measurement and Collection

All measurements are conducted indoors under controlled conditions, with individual testing in which only one participant's comfort perception feedback is assessed at a time. The room temperature is maintained at 23 ± 2 °C, the relative humidity at 40–50%, and the noise levels at 30–40 dB. Before measurement begins, participants are instructed to sit quietly for at least 15 min to relax and to become familiar with the experimental procedures. To avoid spatiotemporal summation of discomfort perception, the interval between bilateral measurements for each participant is set to a minimum of 24 h. The experimental results are recorded by Researcher A and verified by Researcher B.

The measurement duration for each numbered ATEMC measurement tool ranges from 3 to 15 min. The interval between two consecutive measurements is at least 10 min, or longer if the participant indicates a need for additional recovery before continuing. If a participant is unable to differentiate the comfort or discomfort levels among multiple numbered tools, the same comfort or discomfort rating can be assigned to those tools.

To ensure the highest possible measurement accuracy, the following fitting rules are adopted to align the angular reference point of the ATEMC measurement tool with that of the participant's ear. First, the terminal point of the concha side of the ATEMC measurement tool is fitted to the medial side of the medial conchal midpoint. Second, the concha side of the ATEMC measurement tool is aligned parallel to both the participant's ear horizontal plane and the floor of the cavum concha, which is the portion of the cavum concha located between the medial conchal midpoint and the posterior entrance point of the external acoustic meatus. Third, the tragus side of the ATEMC measurement tool is fitted in full contact with the tragion as closely as possible. Finally, all ATEMC measurement tools used by the same participant are fitted to the identical location of that participant's external ear (Fig. 5.14).

To eliminate potential subjective bias in the experimental outcomes, a single-blind procedure is adopted. Specifically, only the numbering information of the measurement tools is disclosed to participants, whereas the actual tools and their

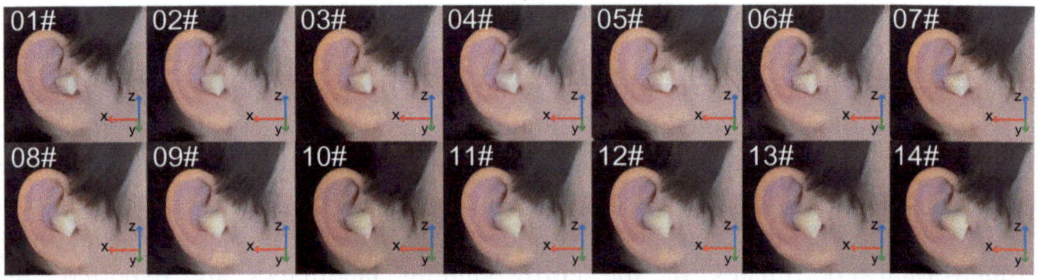

Fig. 5.14 Measuring process of ATEMC prototypes (14# is the extreme discomfort angle rated by the example subject)

dimensions are not revealed. With the assistance of the researchers, participants sequentially use the ATEMC measurement tools. Before each measurement, participants are informed of the tool's number. Each participant is required to designate at least one numbered tool as the "optimal angle." Participants may passively compare the optimal-angle tool with their previous product-use experiences. Therefore, participants are permitted to assign a comfort score to the optimal-angle tool on the basis of such passive comparisons, which means that the perceived optimal comfort level does not necessarily have to be rated as complete comfort.

The duration of a unilateral measurement session for each participant is approximately 3.5–4 h. The procedure is illustrated in Fig. 5.15 and consists of the following steps. First, with the assistance of the researchers, participants are instructed to sequentially use tools 1# through 20#, and the process is terminated if any tool induces extreme discomfort. Second, if none of the first 20 tools induces extreme discomfort, an additional 10 tools with larger angles are fabricated for the participant, and a follow-up session is scheduled, in which the data from the current session are discarded. Third, after each single measurement, or at any point during continuous measurements, participants can assign a comfort rating to the corresponding tool. To determine whether the tragus, the entrance of the external acoustic meatus, and the concha are interrelated in discomfort perception, participants are required to identify the first tool number that elicits discomfort in the entrance and the concha, and to provide the corresponding discomfort ratings. The researchers document participants' descriptions of comfort or discomfort sensations at the tragus, the entrance of the external acoustic meatus, and the concha. If participants report fatigue sensations or show signs of fatigue in their external ear at any point, the experiment is immediately terminated, and the recorded data are discarded.

5.4.3 Validation for the Optimal Angle

To provide valid parameter references and design recommendations, and to verify participants' judgments regarding optimal comfort, a validation test is conducted on the tools that are identified as representing the optimal angle. A total of 92

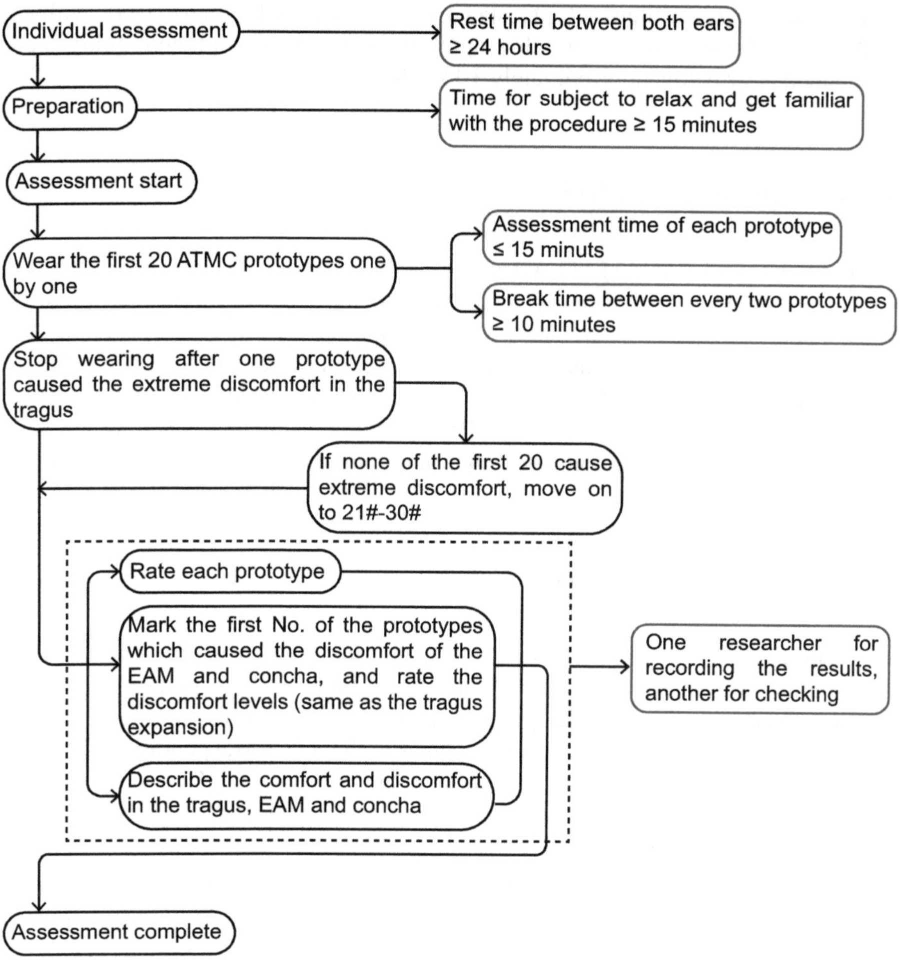

Fig. 5.15 Measuring steps

participants consent to join this additional test, comprising 44 males and 48 females, aged 21–49 years. This validation test also employs a single-blind procedure. For each participant, the ATEMC measurement tool that corresponds to the chosen optimal angle is remanufactured. During testing, participants are not informed of any details regarding the tool being used, and they are not told that both test sessions employ the same tool until completion.

The validation test is conducted during daytime between May and June. Each participant is instructed to wear the ATEMC measurement tool provided bilaterally and to complete two tasks: 0.5 h of outdoor walking and 0.5 h of outdoor jogging. The interval between the two tasks is at least 24 h. The termination condition for an individual test is either the occurrence of significant discomfort in the external ear or the dislodgement of the tool within the prescribed testing duration. Participants monitor and regulate their activity speed using a wearable smart band, with the walking speed maintained at or below approximately 5 km/h and the jogging speed maintained at or below approximately 10 km/h.

5.4.4 *Statistical Analysis*

Data are analysed using MATLAB with a 95% confidence level ($\alpha = 0.05$). The results are considered statistically significant when $p < 0.05$. Before measurement, independent-sample t-tests are applied to examine sex differences in TEI and TEA values across all ATEMC measurement tools. After measurement, one-way ANOVA is employed to test differences in TEI values across age-group distributions under the condition of equal comfort values. Paired-sample t-tests are conducted to analyse bilateral differences in comfort values and in their corresponding TEI values.

5.5 Expansion Analysis

The demographic information of the participants and the measurement outcomes of the auxiliary variable ED-P are summarized in Table 5.6. No participant reports fatigue during the experimental process. In total, 9328 comfort-rating scores are recorded (male: 5058; female: 4270), together with 9328 corresponding TEI values (Tables 5.7 and 5.8; Fig. 5.16).

Table 5.6 Demographics of subjects

Age group	Male N		Age (year)	Height (m)	Weight (kg)	Ear N		ED-P (mm)
20s	50	Mean	25.33	1.77	68.50	100	Mean	10.71
		SD	2.10	0.06	9.48		SD	0.41
30s	50	Mean	35.09	1.75	73.68	100	Mean	10.91
		SD	2.43	0.05	7.03		SD	0.67
40s	50	Mean	45.34	1.74	76.80	100	Mean	11.45
		SD	2.39	0.04	7.33		SD	0.41
50s	50	Mean	54.52	1.72	75.82	100	Mean	11.77
		SD	2.78	0.05	8.55		SD	0.53
Age group	Female N		Age (year)	Height (m)	Weight (kg)	Ear N		ED-P (mm)
20s	50	Mean	24.94	1.68	53.10	100	Mean	10.90
		SD	2.48	0.05	7.38		SD	0.55
30s	50	Mean	34.75	1.68	57.63	100	Mean	11.13
		SD	3.95	0.06	8.87		SD	0.65
40s	50	Mean	45.09	1.65	56.64	100	Mean	11.45
		SD	3.33	0.05	8.26		SD	0.79
50s	50	Mean	55.62	1.62	54.53	100	Mean	11.31
		SD	2.90	0.06	7.18		SD	0.76

Table 5.7 Distribution of ears based on rating scale and ATEMC prototypes (male)

Comfort-discomfort rating		ATEMC tools (male ear, N)																Total
		1#	2#	3#	4#	5#	6#	7#	8#	9#	10#	11#	12#	13#	14#	15#	16#	
Extreme comfort	5		52	12														64
Very high comfort	4.5		58	4														62
	4	7	141	115	9													272
High comfort	3.5	41	18	138	11													208
	3	153	109	111	68	15												456
Moderate comfort	2.5	83	22	18	4	4												131
	2	54		2	294	64	13											427
Little comfort	1.5	46					3											49
	1	15			14	297	64	8										398
Just noticeable comfort	0.5	1				4	54											59
Normal (not sure)	0					4	53	16										73
Just noticeable discomfort	−0.5					12	201	163	24									400
Little discomfort	−1						12	201	163	24								400
	−1.5							8	75	70	20							173
Moderate discomfort	−2							4	134	168	74	20						400
	−2.5								4	85	71	57	20					237
High discomfort	−3									53	177	93	57	20				400
	−3.5										24	38	16	22	16			116
Very high discomfort	−4										34	136	93	40	22	8		333
	−4.5																	0
Extreme discomfort	−5											56	158	104	44	30	8	400
Total		400	400	400	400	400	400	400	400	400	400	400	344	186	82	38	8	5058

Table 5.8 Distribution of ears based on rating scale and ATEMC prototypes (female)

Comfort-discomfort rating		ATEMC tools (Female ear, N))													Total
		1#	2#	3#	4#	5#	6#	7#	8#	9#	10#	11#	12#	13#	
Extreme comfort	5		11	17											28
Very high comfort	4.5		17	6											23
	4	1	36	157	13										207
High comfort	3.5	10	18	88	9										125
	3	45	196	112	50	7									410
Moderate comfort	2.5	7	33	10											50
	2	55	85	9	203	25	7								384
Little comfort	1.5	91	4												95
	1	96		1	123	183	3								406
Just noticeable comfort	0.5	95				91	16								202
Normal (not sure)	0				1	28	50	7							86
Just noticeable discomfort	−0.5				1	65	258	69	7						400
Little discomfort	−1					1	64	181	63	7					316
	−1.5							12	1	4					17
Moderate discomfort	−2						2	129	198	60	11				400
	−2.5								27	25	25				77
High discomfort	−3							2	98	190	57	36			383
	−3.5										3	7			10
Very high discomfort	−4								6	78	112	23	32		251
	−4.5														0
Extreme discomfort	−5									36	156	142	34	32	400
Total		400	400	400	400	400	400	400	400	400	364	208	66	32	4270

5.5.1 Difference Analysis

As shown in Table 5.9, significant sex-related differences ($p < 0.05$) are identified between the participants' measured $ATEMC_{ear}$ values and the rounded design values of the corresponding tool 1# ($ATEMC_{wear}$). Table 5.10 reports the precomputed TEI values, which are provided to all participants, for all ATEMC measurement tools and their sex differences. With the exception of tools 1# and 2#, significant sex differences ($p < 0.05$) are found in the TEI values of all subsequent tools, and the mean TEI values are consistently higher for female participants. This indicates that, beginning with tool 2#, each 1° increment in TEA is associated with significantly greater structural expansion of the auricle-EAM transitional structure and a higher perceptual load in female participants compared with male participants. Consequently, the

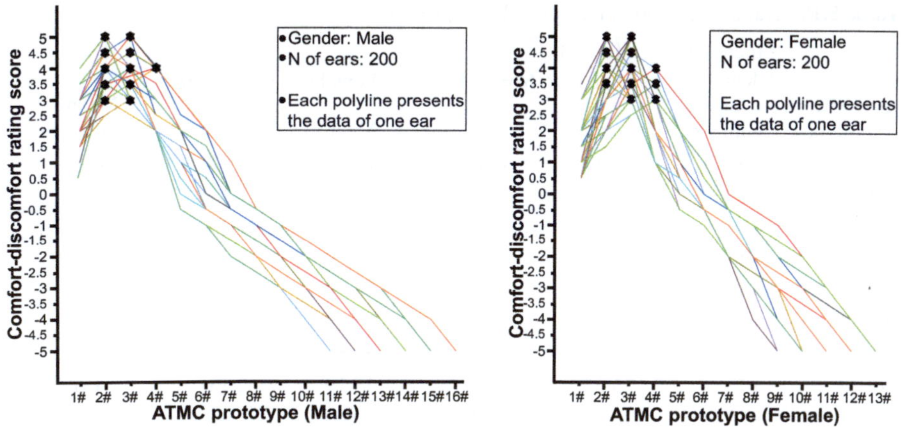

Fig. 5.16 Comfort-discomfort rating scores across ATEMC prototypes

Table 5.9 ATEMC measured values and 1# design values

Item	ATEMC (°)								Mean different
	Male (ear, N = 400)				Female (ear, N = 400)				
	Min	Max	Mean	SD	Min	Max	Mean	SD	
Measured values of subjects	25.03	36.89	29.69	5.68	25.03	34.76	27.88	4.44	1.81[a]
Design values of 1#	26	37	30.22	5.69	26	35	28.32	4.57	1.90[a]

$\alpha = 0.05$
[a] $p < 0.05$ is significant (2-tailed, independent-samples t-test)

measurement data of female and male participants are not pooled for analysis. No significant bilateral differences are identified in the comfort-rating scores or in their corresponding TEI values ($p = 0.869$ and 0.899 for the comfort-rating scores of male and female participants; $p = 0.993$ and 0.995 for the corresponding TEI values). Similarly, the TEI distributions across age groups under the condition of equal comfort values show no significant sex-related differences ($p = 0.08$–0.98). Therefore, no further analysis of bilateral or age-group differences is conducted.

5.5.2 *Functional-Dimension Control Stage of Comfort Perception*

When assigning comfort ratings to ATEMC measurement tools with numbers prior to the optimal angle, no participant reports discomfort at the entrance of the external acoustic meatus or the concha. Likewise, no participant marks tool 1# as the optimal

Table 5.10 Tragus expansion index of ATEMC prototypes

ATEMC prototypes	Increased degree compared with 1#	TEI (%)				Mean different
		Male (ear, N = 400)		Female (ear, N = 400)		
		Mean	SD	Mean	SD	
1#	+0°	1.82	0.86	1.85	1.07	−0.04
2#	+1°	5.21	0.98	5.47	1.15	−0.26
3#	+2°	8.59	1.20	9.09	1.26	−0.51[a]
4#	+3°	12.00	1.46	12.70	1.42	−0.70[a]
5#	+4°	15.38	1.74	16.32	1.59	−0.94[a]
6#	+5°	18.77	2.05	19.94	1.77	−1.17[a]
7#	+6°	22.15	2.35	23.56	1.97	−1.41[a]
8#	+7°	25.55	2.68	27.19	2.18	−1.65[a]
9#	+8°	28.93	2.99	30.81	2.40	−1.88[a]
10#	+9°	32.31	3.32	34.43	2.61	−2.12[a]
11#	+10°	35.70	3.65	38.06	2.84	−2.35[a]
12#	+11°	39.09	3.98	41.68	3.06	−2.59[a]
13#	+12°	42.47	4.31	45.30	3.29	−2.83[a]
14#	+13°	45.86	4.64	48.92	3.51	−3.06[a]
15#	+14°	49.24	4.98	52.54	3.74	−3.30[a]
16#	+15°	52.63	5.31	56.16	3.97	−3.54[a]
17#	+16°	56.01	5.65	59.78	4.20	−3.77[a]
18#	+17°	59.40	5.98	63.41	4.43	−4.01[a]
19#	+18°	62.78	6.32	67.03	4.66	−4.24[a]
20#	+19°	66.17	6.65	70.65	4.90	−4.48[a]

$\alpha = 0.05$

[a] $p < 0.05$ is significant (2-tailed, independent-samples t-test)

angle. Although the $ATEMC_{wear}$ value of tool 1# is close to each participant's own measured value, participants report that tool 1# does not provide sufficient wearing sensation and express concern that it might dislodge at any time.

In the measurement results obtained prior to the optimal angle, the participants' comfort ratings exhibit a monotonically increasing trend (male TEA [TEI] mean = 2.35° [6.27%]; female TEA [TEI] mean = 2.88° [8.56%], Table 5.11, Fig. 5.17). Within this ascending interval of TEA comfort levels, no participant experiences pain or other bodily stimuli attributable to fitting with the ATEMC measurement tools. Because the design values of the tools are close to the participants' measured values and the variations in TEA are relatively small, these results indicate that the participants' wearing comfort perception is fully offset by their functional comfort perception. Thus, the comfort feedback in this interval belongs to functional comfort feedback, and the improvement in comfort levels that corresponds to incremental increases in TEA during this interval is controlled by functional-dimension perception feedback.

Table 5.11 Tragus expansion index and comfort rating score of optimal angles selected by subjects

Gender	Ratings											
	3 (high comfort)				3.5 (high comfort)				4 (very high comfort)			
	N of ears		TEI (%)	TEA (°)	N of ears		TEI (%)	TEA (°)	N of ears		TEI (%)	TEA (°)
Male	40	Mean	6.60	2.55	96	Mean	7.52	2.90	142	Mean	5.51	2.10
		SD	1.79	0.50		SD	0.22	0.31		SD	0.79	0.32
Female	88	Mean	9.81	3.10	92	Mean	9.14	2.89	169	Mean	7.95	2.88
		SD	0.93	0.34		SD	0.76	0.40		SD	1.04	0.44
Gender	Ratings											
	4.5 (very high comfort)				5 (extreme comfort)				Total			
	N of ears		TEI (%)	TEA (°)	N of ears		TEI (%)	TEA (°)	N of ears		TEI (%)	TEA (°)
Male	58	Mean	6.12	2.07	64	Mean	6.03	2.19	400	Mean	6.27	2.35
		SD	0.44	0.26		SD	0.23	0.39		SD	1.09	0.48
Female	23	Mean	7.28	2.26	28	Mean	7.51	2.61	400	Mean	8.56	2.88
		SD	0.54	0.45		SD	0.30	0.50		SD	1.24	0.46

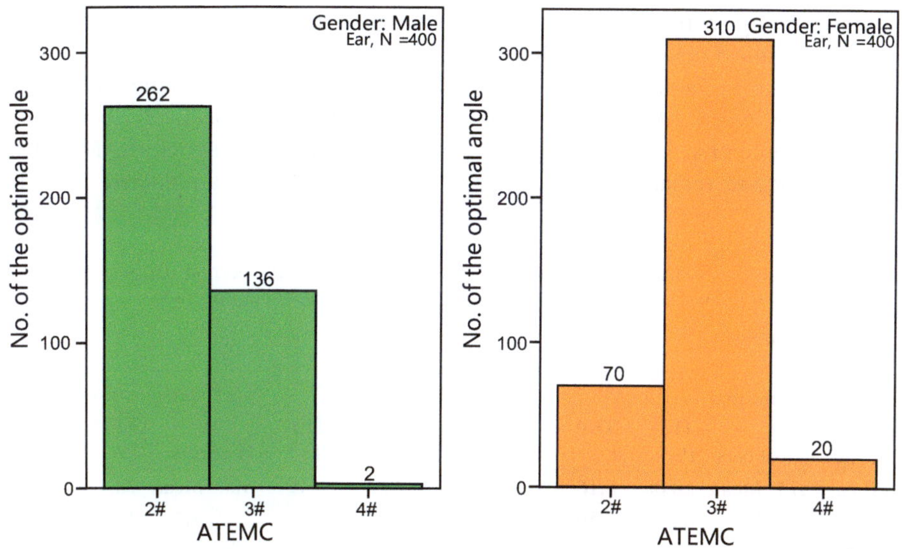

Fig. 5.17 ATEMC prototypes for which the optimal angle exists

5.5.3 Mixed-Dimension Control Stage of Comfort Perception

Each participant identifies a unique optimal angle for the unilateral external ear. In Fig. 5.16, the maximum comfort value of each polyline, represented by a black dot, corresponds to the participant's selected optimal angle. For both male and female participants, the optimal angle falls within the range of tools 2# to 4#. According to the data reported in Fig. 5.17, 99.5% of male participants identify tools 2# or 3# as the optimal angle, whereas 77.5% of female participants identify tool 3#. According to the participants' descriptions of their choices, the tool corresponding to the optimal angle provides sufficient wearing sensation without generating concern about dislodgement. These results suggest that the selection of the optimal angle is influenced by the participants' functional comfort perception feedback.

However, substantial variability is identified among the participants in the comfort-rating scores that are assigned to the optimal angle. The distribution ranges from 3, which represents high comfort, to 5, which represents complete comfort (Table 5.11). Among these ratings, the highest frequency corresponds to "4 (very high comfort)," with 311 ears; the lowest frequency corresponds to "4.5 (very high comfort)," with 81 ears; and only 92 ears are rated as "5 (extreme comfort)." The majority of participants do not assign a score of complete comfort to their optimal angle. The primary reason provided is that, relative to the selected optimal angle, slightly increasing or decreasing the design value of the corresponding measurement tool results in a more comfortable experience. The participants' statements suggest that the external stimulation caused by design-size differences is one of the key considerations in withholding the score of complete comfort. Accordingly, the comfort-rating scores of the optimal angle are also influenced by the participants' wearing comfort perception feedback.

As shown in Table 5.12, the mean values of TEA (TEI) for male and female participants reach 6.61° (18.08%) and 5.73° (17.64%), respectively, before the comfort-perception feedback regarding the expansion of the auricle-EAM transitional structure remains positive. In the measurement results obtained after the

Table 5.12 Tragus expansion index of discomfort rating score for normal (0), just noticeable discomfort (−0.5), moderate discomfort (−2, −2.5) and extreme discomfort (−5)

Gender	Ratings											
	0 (normal)				−0.5 (just noticeable discomfort)				−2 (moderate discomfort)			
	N of ears		TEI (%)	TEA (°)	N of ears		TEI (%)	TEA (°)	N of ears		TEI (%)	TEA (°)
Male	73	Mean	18.08	6.16	400	Mean	20.26	6.49	400	Mean	28.44	8.93
		SD	0.56	0.50		SD	0.94	0.66		SD	1.34	0.87
Female	86	Mean	17.64	5.73	400	Mean	19.85	6.04	400	Mean	26.42	7.87
		SD	0.46	0.62		SD	1.04	0.64		SD	1.53	0.77

Gender	Ratings							
	−2.5 (moderate discomfort)				−5 (extreme discomfort)			
	N of ears		TEI (%)	TEA (°)	N of ears		TEI (%)	TEA (°)
Male	237	Mean	31.10	10.02	400	Mean	41.46	12.64
		SD	0.91	1.00		SD	1.11	1.19
Female	77	Mean	29.52	8.97	400	Mean	37.26	10.66
		SD	2.14	0.83		SD	2.17	1.03

optimal angle, the participants' comfort-rating scores decrease as TEA (TEI) values increase. The findings indicate that when the mean TEA (TEI) values of male and female participants are within the ranges of 2.35° (6.27%) to 6.61° (18.08%) and 2.88° (8.56%) to 5.73° (17.64%), respectively, their sensitivity to external stimulation that is caused by design-size differences progressively increases. Thus, within the declining interval of TEA comfort levels, the participants' wearing comfort perception is partially offset by their functional comfort perception. As TEA (TEI) values increase further, the offsetting effect of functional comfort perception on wearing comfort perception gradually diminishes. Accordingly, on the basis of the overall outcomes of this subsection, the decrease in comfort levels with increasing TEA is jointly governed by functional-dimension perception feedback and physical-dimension perception feedback. In addition, the inability of the participants to clearly distinguish between the comfort-rating scores of two consecutive tools is primarily observed in the vicinity of the optimal angle (Fig. 5.17).

5.5.4 *Physical-Dimension Control Stage of Discomfort Perception*

During unilateral measurement for each participant, one tool number is rated as "−0.5 (just noticeable discomfort)," one as "−2 (moderate discomfort)," and one as "−5 (extreme discomfort)" (Tables 5.7 and 5.8). Out of 400 ears, only 159 yield a score of "0 (normal)." None of the participants assigns a score of "−4 (very high discomfort)" to any tool. When the TEA (TEI) values of male and female participants exceed 6.61° (18.08%) and 5.73° (17.64%), respectively, no indistinguishable comfort-rating scores across tool numbers are observed. For clarity in subsequent discussion, TEI (score) is used to denote the TEI value corresponding to a given comfort-rating score. For example, TEI (−0.5) represents the TEI value at the comfort-rating score of −0.5.

All participants' "extreme discomfort" scores are obtained before TEA reaches 17° (Tables 5.7, 5.8, and 5.12). The tool numbers corresponding to the TEI (−5) values of male and female participants range from tools 11# to 16# and 9# to 13#, respectively. The first reports of tragal discomfort occur when participants use tools corresponding to TEI (−0.5). Although the participants are unable to articulate precise discomfort ratings, they provide vivid descriptions of their sensations, which include the following:

(a) A clear feeling of the tragus being pushed outward by the measurement tool;
(b) A sensation that the tragus is being pulled outward by another person.

Participants report that discomfort at the entrance of the external acoustic meatus begins to emerge when they use measurement tools corresponding to TEI(−2) or TEI(−2.5). However, they are still unable to provide precise ratings of discomfort, acknowledging only the sensation of discomfort. Among all participants, 173 (male:

152; female: 21) identify TEI(−2.5) as the first value associated with discomfort at the entrance. Feedback on discomfort at the entrance can be summarized as follows:

(a) Discomfort originates from, or radiates from, the discomfort of the tragus (TEI(−2));
(b) A distinct sensation of swelling occurs at the entrance of the external acoustic meatus (TEI(−2.5)).

According to the participants' feedback, measurement tools corresponding to TEI(−5) elicit clear discomfort in the concha. Again, this discomfort cannot be precisely scored. Feedback on the perception of extreme discomfort can be summarized as follows:

(a) The tragus is pushed inward by the measurement tool;
(b) Discomfort in the concha spreads to the mastoid skin and to the entire auricle except the lobule;
(c) Pain is perceived at the tragus and at the entrance of the external acoustic meatus;
(d) Discomfort radiates from the tragus to the skin over the mandible and the temporal bone.

As shown in Fig. 4.17, the degree of discomfort that is reported by the participants increases progressively as TEA (TEI) values rise.

Since the participants already provide clear discomfort feedback beginning at TEI(0), the results of this subsection suggest the following: the offsetting effect of functional comfort perception on wearing comfort perception disappears at TEI(0). Therefore, within the discomfort interval of TEA, discomfort increases progressively with rising TEA (TEI) values, with TEI(−5) identified as the exposure limit for TEI. In this interval, the increase in discomfort with larger TEA angles is governed by the participants' physical-dimension perception feedback.

5.5.5 *Dislodgement-Resistance Test of the Optimal Expansion Angle*

During the two validation tasks, which consist of 0.5 h of outdoor walking and 0.5 h of outdoor jogging, the participants report no significant discomfort. The results show that during continuous half-hour fitting, the optimal-angle measurement tool provides a positive comfort experience. No dislodgement of the optimal-angle tool occurs during outdoor walking.

During outdoor jogging, 56 participants completed the 30-minute test without tool dislodgement. In the remaining 36 participants, unilateral or bilateral dislodgement occurs between the 18th and 27th minutes. The primary causes of dislodgement are variations in environmental ergonomic factors, which include slippage that is induced by sweat-related moisture in the external ear and vibration-induced dislodgement that is caused by head movements during running. These results also confirm that the selection of the optimal angle is influenced by the participants' functional comfort perception.

5.6 Implications for Product Design

The auricle-EAM transitional structure, quantified by the ATEMC variable, exhibits both phasic age-growth and persistent age-growth tendencies. Specifically, ATEMC increases monotonically with age, and statistically significant differences are identified across age groups. These results are consistent with earlier research on auricular dimensions, which indicates that auricular width and conchal depth tend to enlarge progressively with age due to cartilage elasticity and soft tissue remodelling. Similarly, prior investigations of the external acoustic meatus suggest that entrance dimensions expand with aging, although the magnitude of change is less pronounced than that observed in auricular structures. The present findings extend this knowledge by illustrating that the transitional structure follows a phasic age-growth pattern, with measurable increments in ATEMC across decades, yet non-significant increases after the age of 69. This trajectory underscores the role of aging in modifying the baseline expansion potential of the transitional structure, which in turn affects the user's physical comfort perception under intrusive fitting.

Sex-related differences are also identified, with male participants exhibiting slightly higher ATEMC values compared with female participants. However, after adjustment for body height and weight, these differences are no longer significant, which suggests that the influence of sex is mediated by body size rather than by intrinsic anatomical distinctions. This conclusion aligns with prior anthropometric studies that report sex differences in ear dimensions but emphasize the necessity of covariate adjustment. The absence of bilateral differences further supports the conclusion that ATEMC is a symmetric variable, which reinforces its reliability as a design reference.

The validation of the optimal angle through walking and jogging tasks provides further evidence that functional comfort influences stability under dynamic conditions. The finding that dislodgement occurs in approximately 36 participants during jogging highlights the role of environmental and activity-related factors in modulating comfort perception. These results echo earlier studies on earmuff and earplug retention, which emphasize the influence of moisture, vibration, and head movement on device stability. The present study therefore bridges the gap between static anthropometric analysis and dynamic ergonomic performance.

The ergonomic implications of the present findings are substantial. First, the quantification of ATEMC and its growth trajectory enables the development of population-specific design standards. For instance, devices that are intended for older populations should account for the greater baseline expansion of the transitional structure, which may reduce tolerance for additional expansion imposed by intrusive devices. Similarly, the identification of sex-related differences underscores the importance of considering body size in design specifications. Such adjustments are critical in minimizing the risk of discomfort and in promoting long-term compliance. Second, the thresholds of TEI that correspond to different levels of comfort provide actionable criteria for product designers. Specifically, intrusive devices should be designed to induce TEI values within the optimal range of 6–18%, thereby maximizing comfort while ensuring stability. Exceeding these thresholds risks eliciting discomfort that compromises usability. The incorporation of these quantitative criteria into design practice can significantly reduce reliance on iterative prototyping and subjective user testing. Third, the staged model of comfort perception suggests that design optimization must balance functional and physical dimensions. At minimal expansion, devices risk instability and dislodgement, whereas at excessive expansion, physical discomfort predominates. Therefore, the challenge for designers is to identify the "sweet spot" where functional stability is achieved without inducing significant discomfort. The ATEMC measurement tools that are developed in this study provide a methodological blueprint for evaluating such trade-offs in the design process.

The multidimensional definition of comfort that is adopted resonates with prior conceptualizations that identify physical, psychological, physiological, and environmental dimensions. However, by introducing the functional dimension as a distinct category, the study extends the existing taxonomy of comfort. The functional dimension is particularly relevant for intrusive ear-related wearables, where retention and intrusiveness constitute primary determinants of user acceptance. This addition also aligns with recent literature on passenger comfort in transportation, which highlights the role of functional constraints in shaping global comfort experience. Thus, the integration of the functional dimension into the comfort framework enriches the theoretical understanding of comfort and provides a more comprehensive basis for ergonomic evaluation.

Competing Interests This study is by "the Fundamental Research Funds for the Central Universities" (NPU Grant No. D5000250035, SEU Grant No. 2242024S30013). Additional support comes from the Northwestern Polytechnical University. The authors have no conflicts of interest to declare that are relevant to the content of this chapter.

References

1. Wang, M., Fan, H., Yu, S., et al. (2022). Effects of variations in the tragus expansion angle on physical comfort for in-ear wearables. *Ergonomics, 65*(10), 1352–1372. https://doi.org/10.1080/00140139.2022.2032377
2. Fan, H., Wang, M., Adamson, C. H., et al. (2020). Anthropometric analysis of external acoustic meatus for ergonomic design. *Proceedings of the Human Factors and Ergonomics Society Annual Meeting, 64*(1), 1833–1835. https://doi.org/10.1177/1071181320641441
3. Doutres, O., Sgard, F. C., Benacchio, S., et al. (2018). Earplug comfort: From subjective assessment on the field to objective measurement and simulation using augmented artificial heads. *The Journal of the Acoustical Society of America, 143*(3), 1910–1930. https://doi.org/10.1121/1.5036223
4. Doutres, O., Sgard, F., Terroir, J., et al. (2019). A critical review of the literature on comfort of hearing protection devices: Definition of comfort and identification of its main attributes for ear-

plug types. *International Journal of Audiology, 58*(12), 824–833. https://doi.org/10.1080/14992027.2019.1646930

5. Wang, M., Fan, H., Harris, C., et al. (2019). Anthropometric characteristics of Chinese auricles for ergonomic design. *Proceedings of the Human Factors and Ergonomics Society Annual Meeting, 63*(1), 1962–1963. https://doi.org/10.1177/1071181319631397
6. Doutres, O., Sgard, F., Terroir, J., et al. (2020). A critical review of the literature on comfort of hearing protection devices: Analysis of the comfort measurement variability. *International Journal of Occupational Safety and Ergonomics*. https://doi.org/10.1080/10803548.2020.1772546
7. Gerges, S. N. Y. (2010). Plenary lectures hearing protectors: Noise attenuation and comfort. *Internoise, 2010*, 1–19.
8. Brown-Rothwell, D. J. (1986). *The comfort of earplugs: Basis for a descriptive model (Master's thesis)*. University of Southampton, Southampton, UK.
9. Casali, J. G., Lam, S. T., & Epps, B. W. (1987). Rating and ranking methods for hearing protector wearability. *Sound & Vibration, 21*(12), 10–19.
10. Hsu, Y. L., Huang, C. C., Yo, C. Y., et al. (2004). Comfort evaluation of hearing protection. *International Journal of Industrial Ergonomics, 33*(6), 543–551. https://doi.org/10.1016/j.ergon.2004.01.001
11. Ahmadpour, N., Lindgaard, G., Robert, J. M., et al. (2014). The thematic structure of passenger comfort experience and its relationship to the context features in the aircraft cabin. *Ergonomics, 57*(6), 801–815. https://doi.org/10.1080/00140139.2014.899632
12. Laitinen, H., & Poulsen, T. (2008). Questionnaire investigation of musicians' use of hearing protectors, self-reported hearing disorders, and their experience of their working environment. *International Journal of Audiology, 47*(4), 160–168. https://doi.org/10.1080/14992020801886770
13. Bockstael, A., Botteldooren, D., De Bruyne, L., et al. (2012). Personal hearing protection and comfort: Indispensable but not a matter of course. In *Proceedings of the European Conference on Noise Control - Euronoise, Prague, Czech Republic*. European Acoustics Association (EAA).
14. Sviech, P. S., Gonçalves, C. G. O., Morata, T. C., et al. (2013). The evaluation of comfort of the personal hearing protection devices as an intervention for hearing loss prevention. *Revista CEFAC, 15*, 1325–1337.
15. Gonçalves, C. G. O., Lüders, D., Guirado, D. S., et al. (2015). Perception of hearing protectors by workers that participate in hearing preservation programs: A preliminary study. *CoDAS, 27*, 309–318. https://doi.org/10.1590/2317-1782/20152014139
16. Terroir, J., Perrin, N., Wild, P., et al. (2021). Assessing the comfort of earplugs: Development and validation of the French version of the COPROD questionnaire. *Ergonomics, 64*(7), 912–925. https://doi.org/10.1080/00140139.2021.1880027
17. Cummins, R. A., & Gullone, E. (2000). Why we should not use 5-point Likert scales: The case for subjective quality of life measurement. In *Proceedings of the second international conference on quality of life in cities* (Vol. 74, pp. 74–93). National University of Singapore.
18. Borg, G. (1990). Psychophysical scaling with applications in physical work and the perception of exertion. *Scandinavian Journal of Work, Environment & Health, 16*, 55–58. https://doi.org/10.5271/sjweh.1815
19. Vink, P., Anjani, S., Udomboonyanupap, S., et al. (2021). Differences and similarities in comfort and discomfort experience in nine countries in Asia, the Americas and Europe. *Ergonomics, 64*(5), 553–570. https://doi.org/10.1080/00140139.2020.1853248
20. Stavrakos, S. K., & Ahmed-Kristensen, S. (2016). Methods of 3D data applications to inform design decisions for physical comfort. *Work, 55*(2), 321–334. https://doi.org/10.3233/wor-162399

Chapter 6
Functional-Dimension Offset Effects on Physical-Dimension Comfort Perception

Abstract Wearing comfort in intrusive external-ear wearables is shaped by the dynamic interaction between functional performance and physical sensation rather than by physical factors alone. Focusing on the auricle-external acoustic meatus (EAM) transitional structure, this chapter employs the Tragus Expansion Index (TEI) to standardize structural expansion and to examine its relationship with comfort perception. Experimental results reveal a bell-shaped exposure-response relationship between TEI and comfort ratings, which is effectively characterized using a Gaussian peak fitting model with high statistical reliability. Analysis indicates a staged perception mechanism across increasing TEI levels. In the comfort-increasing interval, functional-dimension perception, such as stability and wearing assurance, dominates and can fully offset emerging physical discomfort. In the subsequent comfort-decreasing interval, functional and physical dimensions interact, while at higher expansion levels physical discomfort becomes the primary determinant of overall comfort. Discomfort perception is also shown to propagate from the tragus to adjacent auricular and periauricular regions along cutaneous nerve pathways, highlighting its systemic nature. Although sex-related differences diminish after anthropometric adjustment, female participants demonstrate lower discomfort thresholds, which are critical for universal design considerations. Based on quantified comfort thresholds and exposure limits, the study provides ergonomic guidance for balancing functional stability and physical comfort, and advances understanding of perceptual offset effects in the design of intrusive ear-related wearable devices.

Keywords Wearing comfort perception · Tragus expansion index (TEI) · Functional-physical perceptual offset · Exposure-response model · Intrusive ear-wearable design

Comfort perception in ear-related wearables is a multidimensional phenomenon shaped by the interaction between anatomical structures, functional requirements, and users' subjective experience. Traditional ergonomic studies have frequently addressed comfort as a unidimensional construct, focusing primarily on

H. Fan et al., *Engineering Anthropometry of the External Ear for Human-Product Interaction*, https://doi.org/10.1007/978-981-92-2160-8_6

physical discomfort caused by pressure, expansion, or material properties. However, recent developments in wearable design, particularly intrusive external-ear devices, underscore the necessity of recognizing comfort as a complex interplay between physical and functional dimensions. In this context, the present chapter explores how functional-dimension perception can offset or modify physical-dimension perception, thereby influencing users' global comfort judgments.

The auricle-external acoustic meatus (EAM) transitional structure represents a critical interface for intrusive devices such as in-ear headphones, hearing aids, and protective equipment. Expansion of this structure, measured by the tragus expansion index (TEI), provides a quantifiable parameter for analysing the relationship between structural change and comfort perception. Previous research has demonstrated that TEI values correlate with graded comfort-discomfort ratings, producing characteristic bell-shaped exposure-response curves. Building on this foundation, the current study introduces a staged model of perception that distinguishes functional, mixed, and physical dimensions across different ranges of expansion. The methodological framework employs standardized measurement tools (ATEMC tools) and controlled experimental procedures to capture subtle perceptual differences across participants. Gaussian peak fitting (GaussAmp) is used to model the exposure-response relationship, yielding high coefficients of determination and confirming the statistical robustness of the approach. The results reveal that within the comfort-increasing interval of TEI, participants' functional-dimension perception dominates, often masking physical discomfort that might otherwise arise from structural expansion. As expansion progresses into the comfort-decreasing interval, functional and physical dimensions jointly influence perception, while at higher expansion levels, physical discomfort becomes the decisive factor.

A notable finding is the propagation of discomfort across auricular and periauricular regions. Participants frequently describe discomfort that originates at the tragus and radiates toward the concha, the entrance of the EAM, and adjacent skin regions. This pattern is consistent with known pathways of cutaneous innervation and highlights the systemic rather than localized nature of discomfort perception. Furthermore, sex-related analyses demonstrate that while mean ATEMC values do not differ significantly after adjustment for body size, female participants consistently exhibit disadvantaged thresholds in the discomfort range.

The implications for wearable design are substantial. First, the identification of optimal TEI ranges provides concrete guidance for achieving balance between functional stability and physical comfort. Second, specification of exposure limits establishes clear boundaries for safe and acceptable design. Third, the staged perception model emphasizes that intrusive devices should not only minimize physical discomfort but also harness functional-dimension perception to promote stability and user confidence.

6.1 Structural Expansion and the Exposure-Response of Wearing Comfort Perception

6.1.1 Ethics Approval

The studies of this chapter performed in line with the principles of the Declaration of Helsinki. Informed consent was obtained from all participants. The studies of this chapter were approved by the Ethics Committee of Northwestern Polytechnic University (No. IRB No. 201502024, No. 202002024).

6.1.2 Exposure-Response Model

In research and applications oriented towards universal design, it is generally necessary for the results to exhibit strong generalizability and applicability [1]. To this end, this section fits the trends of changes in TEI values and comfort-rating scores across all participants in order to establish the exposure-response relationship between structural expansion of the auricle-EAM transitional structure and changes in comfort-rating scores. For the data trends of each external ear, as TEI increases, comfort-rating scores exhibit a characteristic pattern of first rising and then declining, which presents a curve similar to a bell shape. When the specific comfort-rating scores are regarded as the frequencies at which corresponding TEI values occur, the TEI data of each external ear follow, or approximate, a Gaussian distribution (normal distribution) with an expected value greater than zero (Fig. 6.1).

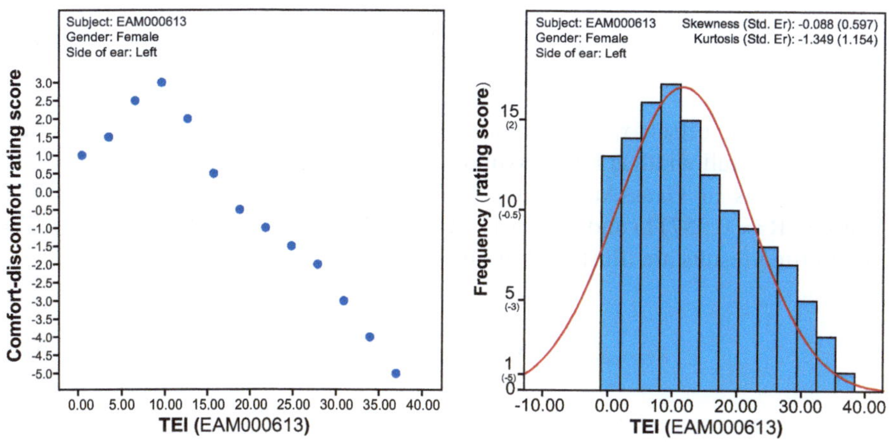

Fig. 6.1 Normal distribution test for TEI with comfort-discomfort rating score as frequency (example)

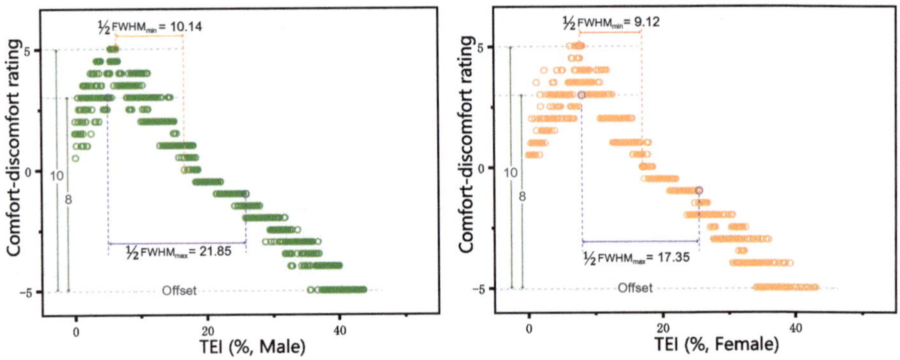

Fig. 6.2 Comfort-discomfort rating scores across Tragus Expansion Index

As shown in Fig. 6.2, all data from the structural-expansion study resemble a set of similar curves. The offset of these curves is −5 (extreme discomfort), and the amplitude ranges from 8 to 10, which corresponds to the optimal comfort values at the optimal angle ranging from 3 (high comfort) to 5 (extreme comfort). The half-width at half-maximum ($\frac{1}{2}FWHM$) of the curves for male participants ranges from 10.14 to 21.85, whereas that for female participants ranges from 9.12 to 17.35. Accordingly, this section employs the amplitude version of the Gaussian peak function (GaussAmp, Eqs. (6.1) and (6.2)) to explore the quantitative relationship between structural expansion of the auricle-EAM transitional structure and changes in comfort-rating scores.

$$y = y_0 + Ae^{-\frac{(x-x_c)^2}{2w^2}} \tag{6.1}$$

$$w = \frac{FWHM}{2\sqrt{\ln 4}} \tag{6.2}$$

where, y_0 is the offset; x_c is the centre point; w is the width; A is the amplitude; $FWHM$ is the full width at half maximum.

The results of GaussAmp fitting are presented in Eq. (6.3) ($R^2 = 0.95714$, Adjusted $R^2 = 0.95711$) and Eq. (6.4) ($R^2 = 0.93767$, Adjusted $R^2 = 0.93763$). The curve-fitting results are illustrated in Fig. 6.3.

$$g_{TEI,male} = -4.97143 + 8.22161e^{-\frac{(TEI-4.13455)^2}{262.74594}} \tag{6.3}$$

$$g_{TEI,female} = -4.75273 + 7.59432e^{-\frac{(TEI-7.83617)^2}{146.11225}} \tag{6.4}$$

where, $g_{TEI,\,male}$ is the mixed wearing-comfort value of male participants with respect to structural expansion of the auricle-EAM transitional structure, $P_{A\,-\,E,}$

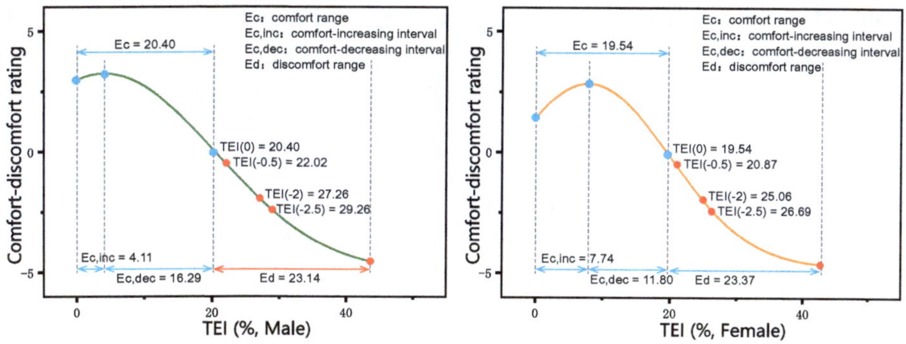

Fig. 6.3 Comfort-discomfort curve fitting of Tragus expansion index employing GaussAmp function

$_{male} \rightarrow -4.97143$, When TEI > 20.40, $g_{TEI,\,male}$ represents the wearing-comfort value associated with physical-dimension perception; $g_{TEI,\,female}$ is the mixed wearing-comfort value of female participants with respect to structural expansion of the auricle-EAM transitional structure, $g_{TEI,female} \rightarrow -4.75273$, When TEI > 19.54, $g_{TEI,female}$ represents the wearing-comfort value associated with physical-dimension perception.

By employing the tragus expansion index (TEI) to standardize and to quantify the degree of expansion of the auricle-EAM transitional structure, the term P_{A-E} can therefore be further expressed using Eq. (3.5), as shown in Eq. (6.5).

$$P_{A-E} = p_{ATEMC} = g_{TEI} \tag{6.5}$$

where, P_{A-E} is the user's comfort perception of the auricle-EAM transitional structural factor; p_{ATEMC} is the user's comfort perception of the quantitative sub-factor within the auricle-EAM transitional structural factor; g_{TEI} is the user's mixed wearing-comfort value with respect to changes in the auricle-EAM transitional structural factor.

As shown in Fig. 6.3, the curve of comfort-rating scores as a function of TEI consists of the comfort range (E_c) and the discomfort range (E_d). The comfort range (E_c) is further divided into the comfort-increasing interval of TEI ($E_{c,\,inc}$) and the comfort-decreasing interval of TEI ($E_{c,\,dec}$). Within the endurance range of comfort perception associated with changes in the tragus expansion index, participants' comfort perception feedback follows a process in which comfort gradually increases, then gradually decreases, and finally transitions into progressive discomfort. The specific composition of the endurance range of comfort perception associated with changes in the tragus expansion index is expressed in Eq. (6.6). Here, $E_{c,\,inc}$ represents the comfort-perception range of the functional dimension, $E_{c,\,dec}$ represents the comfort-perception range of the mixed dimension, and E_d represents the comfort-perception range of the physical dimension.

$$Endurance = E_c + E_d = E_{c,inc} + E_{c,dec} + E_d \tag{6.6}$$

Table 6.1 Wearing comfort thresholds and exposure limits of TEI (male)

(Dis)comfort ratings	Level	(%)
2.73 (functional)	Moderate comfort	0.00 (design value equals the user's measured value)
3.00 (functional)	High comfort	0.11
3.25 (mixed)	Approximate the optimal comfort value	4.11
3.00 (mixed)	High comfort	8.16
2.50 (mixed)	Moderate comfort	11.23
2.00 (mixed)	Moderate comfort	13.44
1.50 (mixed)	Little comfort	15.34
1.00 (mixed)	Little comfort	17.08
0.50 (mixed)	Just noticeable comfort	18.76
0.00 (mixed)	Normal / no discomfort	20.40
−0.50 (mixed)	Just noticeable discomfort	22.02
−1.00 (physical)	Little discomfort	23.69
−1.50 (physical)	Little discomfort	25.41
−2.00 (physical)	Moderate discomfort	27.26
−2.50 (physical)	Moderate discomfort	29.26
−3.00 (physical)	High discomfort	31.53
−3.50 (physical)	High discomfort	34.21
−4.00 (physical)	Very high discomfort	37.63
−4.50 (physical)	Very high discomfort/specified exposure limit	42.91

Tables 6.1 and 6.2 provide the comfort thresholds, exposure limits, and approximate optimal comfort values of TEI for male and female participants, respectively, based on the fitting results. The maximum levels of discomfort (minimum values of the functions) in Eqs. (6.3) and (6.4) are both less than −4.7, which corresponds to very high discomfort, and greater than −5, which corresponds to extreme discomfort. For practical application, this section specifies the exposure limit of TEI uniformly as TEI(−4.5). In other words, the TEI value at −4.5 (very high discomfort) is designated as the specified exposure limit for this section. According to practical requirements, higher specified exposure limits can be calculated using Eqs. (6.3) and (6.4).

After adjustment for height and weight, the sex-related differences of ATEMC are not statistically significant. Therefore, the mean value of ATEMC, together with Tables 6.1 and 6.2, is used to calculate the comfort thresholds, exposure limits, and approximate optimal comfort values of auricle-EAM transitional structure expansion (Table 6.3). When the design value is smaller than the TEI value corresponding to the optimal angle, the likelihood of dislodgement of intrusive ear-related wearables increases considerably. Accordingly, it is not recommended to select values within the $E_{c,\,inc}$ interval. Since the mean value of ATEMC is used as the basis for threshold calculation in this section, the specific recommendations for TEI values in universal-design research or applications across both sexes are provided as follows (Eq. (6.7)):

Table 6.2 Wearing comfort thresholds and exposure limits of TEI (female)

(Dis)comfort ratings	Level	TEI (%)
0.24 (functional)	Just noticeable comfort	0.00 (design value equals the user's measured value)
1.50 (functional)	Little comfort	0.35
2.00 (functional)	Moderate comfort	1.98
2.50 (functional)	Moderate comfort	4.46
2.84 (mixed)	Approximate the optimal comfort value	7.74
2.50 mixed)	Moderate comfort	11.48
2.00 (mixed)	Moderate comfort	13.69
1.50 (mixed)	Little comfort	15.36
1.00 (mixed)	Little comfort	16.84
0.50 (mixed)	Just noticeable comfort	18.21
0.00 (mixed)	Normal/no discomfort	19.54
−0.50 (physical)	Just noticeable discomfort	20.87
−1.00 (physical)	Little discomfort	22.19
−1.50 (physical)	Little discomfort	23.58
−2.00 (physical)	Moderate discomfort	25.06
−2.50 (physical)	Moderate discomfort	26.69
−3.00 (physical)	High discomfort	28.53
−3.50 (physical)	High discomfort	30.78
−4.00 (physical)	Very high discomfort	33.82
−4.50 (physical)	Very high discomfort/specified exposure limit	39.39

- The recommended range of target comfort values for research or application is from 2.5 (moderate comfort) to −4.5 (very high discomfort).
- After selecting a specific target comfort value within the range of 2.5 to −4.5, the TEI value should be chosen according to the sex that exhibits weaker performance in the TEI distribution, and female TEI values must not be selected from within their comfort-increasing interval.

$$TEI_{(\sigma)} = \min\left\{TEI_{male}, TEI_{female}\right\} \tag{6.7}$$

where, $\sigma \in [2.5, -4.5]$, $TEI_{female} \notin E_{c,\,inc}$.

6.1.3 Propagation Pathways of Wearing Discomfort

During changes in TEI, according to participants' descriptions, they perceive that their discomfort originates at a certain part of the external ear and is transmitted, diffused, spread, or radiated to other regions of the external ear or the head-face

Table 6.3 Wearing comfort thresholds and exposure limits of the auricle-EAM transitional structure

(Dis)comfort ratings	ATEMC (°)	
	Male	Female
0.24 (functional)	/	31.66
1.50 (functional)	/	31.77
2.00 (functional)	/	32.29
2.50 (functional)	/	33.07
2.73 (functional)	31.66	/
2.84 (approximate optimal threshold for female participants)	/	34.11
3.00 (functional)	31.69	/
3.25 (approximate optimal threshold for male participant)	32.96	/
3.00 (mixed)	34.24	/
2.50 (mixed)	35.22	35.29
2.00 (mixed)	35.92	35.99
1.50 (mixed)	36.52	36.52
1.00 (mixed)	37.07	36.99
0.50 (mixed)	37.60	37.43
0.00 (mixed)	38.12	37.85
−0.50 (mixed)	38.63	38.27
−1.00 (physical)	39.16	38.69
−1.50 (physical)	39.70	39.13
−2.00 (physical)	40.29	39.59
−2.50 (physical)	40.92	40.11
−3.00 (physical)	41.64	40.69
−3.50 (physical)	42.49	41.40
−4.00 (physical)	43.57	42.37
−4.50 (physical/specified exposure limit)	45.25	44.13

area. The participants' accounts are similar to the phenomenon of "the spread of discomfort perception" reported by Ravnik et al. [2]. In Ravnik et al.'s study, participants reported that neck discomfort caused by prolonged sitting during car driving propagated to the shoulders and lower back. When combined with the intrinsic discomfort of these body parts, the participants' overall comfort values decreased rapidly.

Based on participants' descriptions, the propagation of wearing discomfort induced by structural expansion of the auricle-EAM transitional structure proceeds as follows:

1. TEI(−2): tragus → entrance of the external acoustic meatus;
2. TEI(−5): tragus → skin over the mandibular and temporal bone regions; concha → skin over the mastoid region and the entire auricle except the lobule.

At present, a specific explanation for the cause of this phenomenon cannot be provided. However, in reference to the pain-threshold findings of Fransson-Hall et al. [3], a hypothesis can be proposed: the propagation of discomfort across the skin surface may result from the combined effects of cutaneous nerve distribution, superficial muscle arrangement, and subcutaneous fat filling. Discomfort propagates within the same cutaneous nerve control region, the arrangement of superficial muscles may produce a non-uniform propagation pattern, and the density of subcutaneous fat filling may partly impede propagation. This hypothesis provides a reasonable explanation for why discomfort does not propagate to the lobule, since the lobule contains no cartilage or muscle tissue but instead consists mainly of abundant subcutaneous fat and a small amount of loose connective tissue.

On the basis of the distribution of cutaneous nerve control in the external ear and periauricular regions, the propagation pathways of wearing discomfort can be summarized as follows:

1. Mandibular nerve [V3] pathway: tragus → entrance of the external acoustic meatus, mandibular region, and temporal bone region;
2. Lesser occipital and greater auricular nerve pathway: concha → mastoid region and the entire auricle except the lobule.

In addition, numerous intrinsic and extrinsic muscles are distributed in and around the external ear (Fig. 6.4). Among these, the posterior auricular muscle is known to

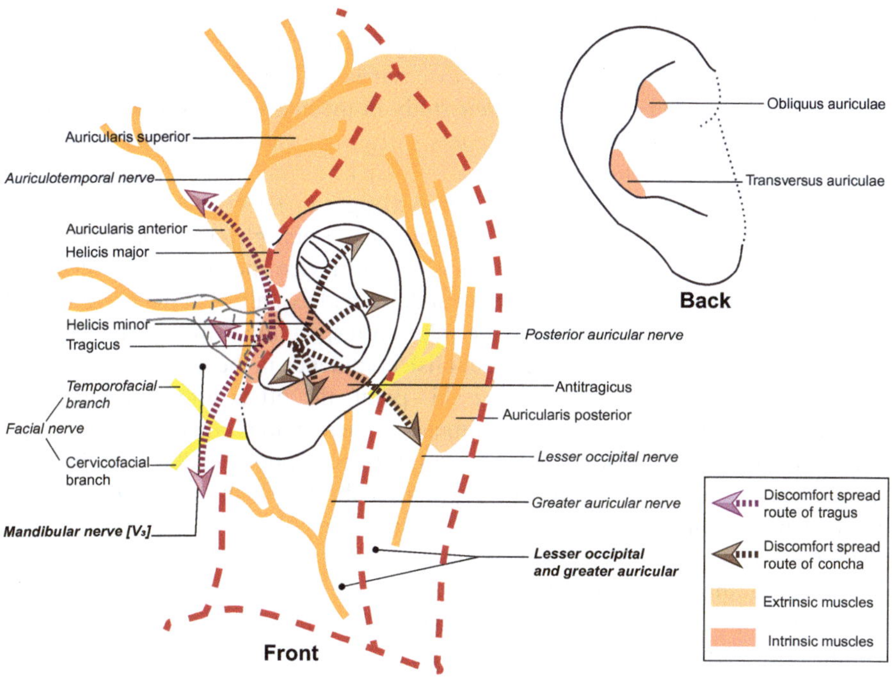

Fig. 6.4 Discomfort spread routes (cutaneous innervation in bold and italic)

be highly sensitive to external stimulation [4, 5]. Furthermore, the skin of the external ear lacks dermal papillae and is extremely sensitive to pain [6]. Therefore, the actual propagation pathways may be more complex. According to the propagation pathways of wearing discomfort induced by structural expansion of the auricle-EAM transitional structure, comfort optimization can be implemented in the overlapping or concentrated regions of the sub-pathways for intrusive ear-related wearable design and application, including:

1. selecting soft design materials;
2. reducing or avoiding structural designs that are in full contact with these regions.

Intrusive design is one of the most feasible universal design approaches for the external-ear wearables. This approach relies on full contact between the device's structure and surface form with the user's tragus, concha, and external acoustic meatus, thereby facilitating auditory enhancement, hearing protection, and functional noise-reduction performance. In particular, for users who are exposed to noise levels that may cause hearing loss or damage, this approach is undoubtedly the optimal solution among limited available options [7].

In a study on tragus expansion, Fan et al. [8] measured participants' perception of tragus-expansion comfort using five intrusive products of identical type but different sizes. They found that as the device size increased, the degree of tragus expansion progressively increased, and participants' perception of tragus expansion shifted from comfort to discomfort. Although the measurement method and variables differed, the results of the comfort-decreasing and discomfort ranges of TEI are consistent with those reported by Fan et al. Because individual participants varied in the structural dimensions of their external ears, functional-dimension perception feedback was not reflected in their study in the absence of standardization of device increments relative to participants' measured values. For this reason, the ATEMC measurement tool and the tragus expansion index are employed in the present study to measure and to standardize changes in TEA.

For the comfort design of intrusive wearables, users' physical comfort perception feedback is established on the basis of the device meeting basic functional requirements. This viewpoint is further validated and explained in the present book. When participants' TEI increases to the value corresponding to the optimal comfort-rating score, they begin to passively establish confidence in the use of the device [9] and a sense of task safety [7]. After this point, the offset effect of functional-dimension perception on physical-dimension perception gradually diminishes. Therefore, the foundation of physical comfort perception feedback in intrusive design lies in functional comfort perception feedback, which is formed only after participants establish confidence and task safety in relation to the device. Once these basic conditions of physical comfort perception feedback in intrusive design are achieved, the degree of expansion of the auricle-EAM transitional structure or the magnitude of the tragus expansion angle directly influences the user's wearing-comfort experience of the external-ear devices.

6.2 Perceptual Offset Effects

6.2.1 General Description and Possible Explanations

Rating scales have been widely applied to measure users' comfort perception in order to obtain feedback information related to physical, physiological, psychological, or environmental aspects that cannot be directly measured. The tragus-expansion comfort scale employed in this book defines 13 comfort levels and 21 comfort scores, which are used to capture participants' perceptual information regarding subtle differences in the degree of structural expansion. The results indicate that although the optimal comfort level corresponding to the optimal angle varies among individuals, the overall trend of TEI within the comfort range is largely consistent. Compared with "no discomfort (0)," participants exhibit higher perceptual sensitivity to "ambiguous discomfort (−0.5)." As TEI gradually increases, participants provide insensitive feedback regarding the differences between "very high discomfort" (−4 to −4.5). Although no participants report obvious fatigue, imperceptible psychological or physiological fatigue may still exist, which could be one of the potential factors influencing perceptual sensitivity. Future research should therefore conduct further evaluation of the influence of fatigue.

The rating-scale results indicate that participants' perception of comfort in functional dimensions exerts an offset effect on their perception of physical comfort. Within the comfort range of TEI, functional comfort perception either fully offsets physical comfort perception (comfort-increasing interval) or partially offsets physical comfort perception (comfort-decreasing interval), which results in the absence of global or wearing discomfort within a certain TEI range. The measurement tools employed are intrusive devices. Even tool 1# maintains full contact with the auricle-EAM transitional structure. In theory, the skin of the external ear should exhibit minor or very minor discomfort due to external stimulation induced by device fitting. However, within a specific TEA range (male: 6.49°; female: 6.04°; Table 4.12), participants do not clearly perceive any physical discomfort induced by skin sensitivity. Moreover, according to the findings of Fan et al. [8], when tragus expansion deviates within 7° from the original position, participants report comfort feedback. The present results are close to that finding.

Possible explanations for the occurrence of perceptual offset effects are as follows. First, the effect may be related to participants' unintentional responses or feedback based on their previous usage experiences with the functional dimension [10]. Second, the effect may be associated with anticipatory judgments derived from passive psychological comparisons [11]. Third, the perceptual intensity of functional comfort perception may exceed that of physical discomfort caused by skin sensitivity. To fully understand the causes of perceptual offset effects, additional comparative and follow-up studies are required. In addition, no participants designate tool 1# as the optimal-angle tool, which suggests that participants do not trust design values close to their actual measured angles in terms of functional assurance. According to participants' descriptions, higher comfort levels are derived from structural expansion that provides a sufficient wearing sensation. Since functional-dimension perception completely offsets physical-dimension perception, it is not recommended to use the measured values or values close to the measured value as the ATEMC design value for wearable devices.

6.2.2 Interactive Associations of Wearing Discomfort among the Tragus, Concha, and Entrance of the External Acoustic Meatus

In the human-device interactions of intrusive external-ear wearables, the tragus, the concha, and the entrance of the external acoustic meatus constitute the primary regions where interaction occurs. Anatomically, the anterior side of the entrance is connected with the medial surface of the tragus, whereas the posterior side is attached to the conchal cartilage. These structural linkages imply that external forces applied to the transitional region inevitably generate combined effects on all three anatomical sites. As evidenced by the data, participants' wearing discomfort (representing their physical discomfort perception) is not confined to a single site but rather exhibits clear interactive associations among the tragus, the concha, and the entrance of the external acoustic meatus.

When the mean amplitude of TEA (TEI) increases to 6.49° (20.26%) in male participants and 6.04° (19.85%) in female participants, participants consistently report ambiguous discomfort (−0.5). At this stage, however, no obvious discomfort feedback is reported in relation to the concha or the entrance of the external acoustic meatus. With further expansion, when the mean amplitude of TEA (TEI) reaches 8.93° (28.44%) for males and 7.87° (26.42%) for females, discomfort begins to emerge at the entrance of the external acoustic meatus. Finally, when participants' perception escalates to "extreme discomfort (−5)," the concha is also reported as uncomfortable. At this point, the discomfort of the concha is primarily attributed to the pushing and compression of the conchal plate by the measurement tool. This sequence of perception indicates a gradual progression in which discomfort originates at the tragus, then extends to the entrance of the external acoustic meatus, and finally involves the concha. A possible explanation is that the tragus, as the most protruding and structurally mobile component of the auricle-EAM transitional structure, responds to mechanical expansion earlier and more intensely than the concha and the entrance. When the tragus is displaced laterally, the force distribution propagates to adjacent connective regions, which indirectly increases the load on the entrance. Once TEA reaches higher amplitudes, the conchal plate, which is relatively rigid, experiences direct mechanical compression, resulting in distinct discomfort.

A comparison of perceptual intensities further confirms this progression. Among the three anatomical regions, participants consistently report the highest discomfort intensity at the tragus, followed by moderate discomfort at the entrance, whereas the concha is perceived as the least sensitive site under equivalent expansion magnitudes. This outcome is consistent with previous findings that the tragus region exhibits the highest sensitivity to external stimulation [12]. Because of this sensitivity hierarchy, participants find it difficult to provide precise comfort-rating scores for the concha and the entrance, as their perceptual feedback is dominated by the stronger discomfort arising from the tragus. These interactive associations suggest that discomfort perception is not merely a localized phenomenon but a systemic outcome of structural coupling. Much like the discomfort feedback observed in prolonged sitting conditions described in Chap. 5, where pressure interaction among multiple anatomical sites amplifies discomfort, the interplay among the measurement tool, the tragus, the concha, and the entrance of the external acoustic meatus collectively escalates discomfort perception. In practical terms, this implies that intrusive wearable devices must be carefully designed to avoid simultaneous pressure on all three regions. Otherwise, the synergistic effects of interaction may amplify discomfort beyond the sum of individual contributions.

From an ergonomic perspective, these findings underscore the importance of focusing on the tragus as the critical entry point of wearing discomfort. While the concha and the entrance eventually contribute to discomfort under larger expansions, the tragus functions as the primary trigger. Therefore, strategies for comfort optimization should prioritize reducing mechanical pressure on the tragus, either by selecting softer materials or by altering geometric configurations that minimize direct expansion forces. The interactive discomfort relationships among the tragus, the concha, and the entrance highlight the necessity of viewing these structures not in isolation but as an integrated system in which mechanical load and perceptual outcomes are tightly interconnected.

6.2.3 Exposure-Response Models

The introduction of TEA and TEI provides a means of normalizing individual variability in ATEMC, thereby allowing the comparison of perceptual responses across participants. The results indicate that TEI values within the range of 6–18% correspond to the interval of optimal comfort perception, whereas values exceeding 30% progressively induce discomfort that culminates in extreme discomfort at approximately 40% expansion. These thresholds represent critical parameters for ergonomic design, as they delineate the boundaries between tolerable and intolerable fitting conditions.

The staged model of comfort perception (functional-dimension control, mixed-dimension control, and physical-dimension control) further clarifies the mechanisms through which expansion influences perception. At low levels of TEA, comfort perception is dominated by functional considerations such as stability and retention. Participants report insufficient wearing sensation at minimal expansion, which reflects the importance of functional comfort in securing device usability. As

TEA increases toward the optimal angle, both functional and physical dimensions contribute, leading to high or very high comfort ratings. Beyond this point, physical discomfort emerges as the dominant dimension, with participants describing pain, swelling, and radiating discomfort across auricular regions. This staged model advances prior theoretical frameworks of comfort, which have often treated comfort as a unidimensional construct, by demonstrating that different dimensions assume control at different levels of structural expansion.

According to the fitting results of the exposure-response model, the endurance range of TEI comfort perception includes the comfort-increasing interval ($E_{c,\ inc}$), the comfort-decreasing interval ($E_{c,\ dec}$), and the discomfort range (E_d). Possible explanations for this outcome include the offset effect of functional-dimension perception on physical-dimension perception, or a potential order effect inherent in the employed methodology. The measurement tools used are non-random with respect to angular variation. Even though the specific information about the measurement tools is not disclosed to participants, they may still perceive the sequence of angular changes through variations in tragus expansion.

Although the sizes of the E_c ranges of TEI (20.40% and 19.54%) differ between sexes, the sizes of the E_d ranges (23.14% and 23.37%) are closer. During the process of comfort decline, however, the ATEMC values of female participants remain disadvantaged. Therefore, when conducting discomfort-optimization research in the E_d range, the ATEMC values of female participants should still serve as the primary data reference.

The exposure-response model presented belongs to the category of statistical models. In future research, its accuracy can be further constrained by incorporating larger participant datasets or by substituting data from target-user populations. For the comfort design of intrusive wearable devices, particularly those intended for sports or outdoor applications, the results of the dislodgement-resistance test suggest that design values should be selected within the $E_{c,\ dec}$ range. The potential risk of dislodgement can also be mitigated by introducing auxiliary components such as ear bands.

Based on the exposure-response model, several design recommendations for TEA or TEI values are provided as follows:

1. For universal-design applications, TEI values for female participants within the range of TEI(2.5) to TEI(0) (11.48%–19.54%, mixed wearing-comfort thresholds) are most appropriate. If the design is targeted only at male users, the recommended range is TEI(3.25) to TEI(0) (14.11%–20.40%, mixed wearing-comfort thresholds).
2. For both universal design and male-specific design, TEI values below 11.23% (corresponding to male TEI(2.5)) are not recommended.
3. TEA values greater than 20.20% (corresponding to male TEI(−0.5)) may induce discomfort in both sexes; therefore, thresholds above 20.20% are not recommended for universal-design applications.

6.3 Limitations of Application

6.3.1 Sample-Related Limitations and Generalizability

The present model is based on a participant population that, while sufficient for statistical analysis within the targeted group, still represents what can be described as a weak sample database in terms of explanatory power. In other words, the overall sample composition allows for reliable description of within-group phenomena but does not yet achieve the level of robustness necessary to support cross-group comparative research or integrative design studies that require strong sample databases.

The findings reported here are closely tied to the demographic characteristics of the recruited participants, who were exclusively drawn from the Chinese population. This focus ensures a certain degree of homogeneity and provides a precise understanding of ear-related anthropometric and comfort characteristics within this specific group. However, it also introduces limitations with respect to generalizability. Comfort thresholds, exposure limits, and perceptual offset effects documented may not necessarily translate to other populations without adjustment. Differences in auricular morphology, cartilage elasticity, and external acoustic meatus dimensions across ethnic groups have been documented in prior anthropometric studies. Consequently, extending the results to populations outside of China without direct validation may introduce inaccuracies or overgeneralizations.

The implication for universal design is clear. Until robust cross-population comparisons are performed, any application of these findings to multi-ethnic or global user groups must be treated with caution. Designers and researchers seeking to incorporate the present results into integrative design frameworks should recognize that the explanatory power is constrained by its within-group focus. While this book contributes valuable insight into the auricle-EAM transitional structure and its relationship with comfort perception, the results cannot yet serve as universal parameters across all populations.

Another notable limitation concerns the age distribution of participants. The exposure-response study did not include individuals in the 15–19 year range or the 60–79 year range. As a result, the general conclusions are derived under the assumption that age-group distributions do not introduce significant differences beyond those captured within the sampled groups. While such an assumption is statistically expedient, it is not empirically validated for the excluded age ranges.

For younger participants (15–19 years), auricular cartilage and external acoustic meatus morphology may still be undergoing subtle developmental changes. Persistent age-growth and phasic age-growth trajectories suggest that transitional structural dimensions may not stabilize until the early twenties. Therefore, excluding this age range limits the ability to address comfort perception in adolescents, a group increasingly targeted by consumer electronics such as wireless earbuds.

For older participants (60–79 years), age-related remodelling of cartilage and soft tissues may alter both baseline auricular dimensions and skin sensitivity. Prior studies indicate that elasticity decreases with age, and conchal depth may expand further in older individuals. Without empirical data from this age group, the current results cannot reliably predict comfort thresholds or exposure limits for elderly users, who may represent a particularly vulnerable population with respect to intrusive device design.

Future research must therefore address these gaps by recruiting participants across the full age spectrum, including both adolescents and older adults. Only through such inclusion can exposure-response models be considered comprehensive. Until then, application of the present findings to the younger and older populations must remain tentative, with recognition of their potential limitations.

6.3.2 Psychological and Physiological Fatigue Effects

Although no participants reported explicit psychological or physiological fatigue during the experimental procedures, it cannot be definitively ruled out that subtle or imperceptible fatigue influenced participants' sensitivity to discomfort. Discomfort perception, especially within the physical dimension, is known to be modulated by central and peripheral fatigue mechanisms. Even in the absence of conscious reporting, low-level fatigue may diminish the acuity of comfort perception or delay the onset of reported discomfort.

The potential interference of fatigue is particularly relevant in studies involving repetitive fitting of intrusive measurement tools. Each session required participants to evaluate multiple ATEMC measurement tools sequentially, and although breaks were provided, the cumulative load of mechanical stimulation may still have produced fatigue effects. This issue warrants careful consideration because the sensitivity of discomfort perception directly influences the precision of exposure-response modelling.

Future research should incorporate dedicated fatigue assessments, including both subjective scales and objective physiological measures. For example, electromyography could be employed to detect subtle muscular fatigue in periauricular regions, while psychophysiological indicators such as heart-rate variability might reveal stress-related responses. Such multi-modal monitoring would clarify whether the absence of reported fatigue genuinely indicates no effect, or whether imperceptible fatigue exerts a hidden influence on discomfort perception.

6.3.3 Applicability to Different Product Categories

The exposure-response model developed is specifically oriented toward intrusive external-ear wearable devices that induce expansion of the auricle-EAM transitional structure by design. Consequently, the findings may not be directly applicable to devices composed primarily of soft materials, such as foam earplugs or adhesive signal patches applied to the ear canal. Soft-material devices do not generate the same structural expansion forces and therefore engage different comfort mechanisms. For example, earplugs made of foam or silicone primarily rely on surface pressure distribution within the ear canal, with minimal involvement of the tragus or concha. Similarly, thin-film patches designed for acoustic or biosignal transmission adhere to the skin rather than expand internal structures. In these cases, comfort perception is governed more by surface adhesion, material compliance, or skin tolerance than by angular expansion. This distinction underscores the importance of recognizing the boundaries of the present results. While this book contributes to the understanding of intrusive-device ergonomics, its direct transfer to soft-material product categories is not warranted. Future research should therefore seek to adapt the exposure-response framework for non-intrusive or soft-intrusive designs, potentially by redefining exposure variables in terms of surface pressure or adhesive force rather than TEI.

The ATEMC measurement tools employed vary systematically in angular increments. This non-random design introduces the possibility of order effects or memory effects in participants' responses. Even though the specific design details of each tool were not disclosed to participants, they may still have inferred the sequence of angular changes through the progressive sensation of tragus expansion. Order effects arise when participants' responses are influenced not solely by the stimulus itself but by its position in a sequence. For example, a tool presented later in a series might be perceived differently simply because it follows tools with smaller or larger angles. Memory effects occur when prior experiences with earlier tools influence judgments of subsequent ones, potentially biasing perception of comfort thresholds. These potential effects pose limitations on the precision of exposure-response modelling. Although these risks could be mitigated through single-blind procedures and inter-session recovery intervals, complete elimination of order and memory effects cannot be guaranteed. Future studies should consider randomized presentation of angular variations or counterbalanced sequences to better isolate true perceptual responses from procedural artifacts.

6.3.4 Weight, Time, and Fatigue Interaction Effects

The relationship between measurement-tool weight and functional-dimension perception feedback was not systematically examined. Although participants did not report discomfort attributable to weight differences (the maximum weight variation among tools was only 1.1 g), this absence of reported effect should not be taken as definitive evidence of no influence. Over extended periods, even minor weight differences may accumulate to produce discomfort, especially when combined with increased TEI values.

The present model focused primarily on short-term exposures within controlled laboratory settings. However, in real-world contexts, intrusive wearables are often used for prolonged durations, such as hours of continuous listening or monitoring.

Under these conditions, weight, time, and fatigue may interact in complex ways. A device that is tolerable for 15 minutes in a laboratory may produce discomfort after 2 h of continuous use, not because of angular expansion alone but because of cumulative load effects.

Future research should therefore extend exposure-response modelling to longer timeframes, incorporating repeated measures of comfort perception across extended usage. Longitudinal designs would reveal whether TEI-related thresholds remain stable over time or whether fatigue and weight effects shift comfort perception progressively. Such insights would be invaluable for designing devices intended for sustained use, particularly in occupational or clinical contexts.

In summary, the limitations of the present model arise from sample composition, age distribution, potential fatigue effects, product-category applicability, methodological considerations, and incomplete examination of weight-time interactions. Each of these limitations constrains the explanatory scope of the exposure-response model and highlights directions for future investigation.

To strengthen generalizability, future studies should recruit broader and more diverse participant samples, including cross-population comparisons and inclusion of both adolescent and elderly groups. To clarify fatigue effects, multimodal assessments should be integrated to capture subtle psychological and physiological influences. To extend applicability, research should adapt exposure-response modelling for soft-material devices and explore context-specific modifications. Methodological rigor should be enhanced through randomization and counterbalancing, while longitudinal designs should address cumulative weight and time effects.

For design practice, these limitations carry important implications. Universal design must recognize the boundaries of current findings, particularly with respect to population differences and usage contexts. At the same time, the exposure-response framework developed here provides a foundation for iterative refinement. By systematically addressing the identified limitations, future research can progress toward a more comprehensive, robust, and universally applicable model of comfort perception for intrusive and non-intrusive external-ear wearables alike.

Competing Interests This study is by "the Fundamental Research Funds for the Central Universities" (NPU Grant No. D5000250035, SEU Grant No. 2242024S30013). Additional support comes from the Northwestern Polytechnical University. The authors have no conflicts of interest to declare that are relevant to the content of this chapter.

References

1. Terroir, J., Perrin, N., Wild, P., et al. (2021). Assessing the comfort of earplugs: Development and validation of the French version of the COPROD questionnaire. *Ergonomics, 64*(7), 912–925. https://doi.org/10.1080/00140139.2021.1880027
2. Ravnik, D., Otáhal, S., & Dodič, F. M. (2008). Using different methods to assess the discomfort during car driving. *Collegium Antropologicum, 32*(1), 267–276.

3. Fransson-Hall, C., & Kilbom, Å. (1993). Sensitivity of the hand to surface pressure. *Applied Ergonomics, 24*(3), 181–189. https://doi.org/10.1016/0003-6870(93)90006-u
4. O'Beirne, G. A., & Patuzzi, R. B. (1999). Basic properties of the sound-evoked post-auricular muscle response (PAMR). *Hearing Research, 138*(1-2), 115–132. https://doi.org/10.1016/s0378-5955(99)00159-8
5. Kızıltan, M. E., Gündüz, A., & Şahin, R. (2010). Auditory evoked blink reflex and posterior auricular muscle response: Observations in patients with HFS and PFS. *Journal of Electromyography and Kinesiology, 20*(3), 508–512. https://doi.org/10.1016/j.jelekin.2009.07.009
6. Standring, S. (2016). *Gray's anatomy: The anatomical basis of clinical practice* (41st ed.). Elsevier.
7. Stavrakos, S. K., Ahmed-Kristensen, S., & Goldman, T. (2016). Using archetypes to create user panels for usability studies: Streamlining focus groups and user studies. *Applied Ergonomics, 56*, 108–116. https://doi.org/10.1016/j.apergo.2016.02.016
8. Fan, H., Yu, S., Chu, J., et al. (2019). Anthropometric characteristics and product categorization of Chinese auricles for ergonomic design. *International Journal of Industrial Ergonomics, 69*(1), 118–141. https://doi.org/10.1016/j.ergon.2018.11.002
9. Fu, F., & Luximon, Y. (2019). Fit and comfort perception on hearing aids: A pilot study. In *International conference on human systems engineering and design: Future trends and applications* (pp. 360–364). Springer. https://doi.org/10.1007/978-3-030-27928-8_55
10. Ahmadpour, N., Robert, J. M., & Lindgaard, G. (2016). Aircraft passenger comfort experience: Underlying factors and differentiation from discomfort. *Applied Ergonomics, 52*, 301–308. https://doi.org/10.1016/j.apergo.2015.07.029
11. Wang, M., Fan, H., Yu, S., et al. (2022). Effects of variations in the tragus expansion angle on physical comfort for in-ear wearables. *Ergonomics, 65*(10), 1–37. https://doi.org/10.1080/00140139.2022.2032377
12. Blondé-Weinmann, C., Joubaud, T., Zimpfer, V., et al. (2021). Characterization of cartilage implication in protected hearing perception during direct vibro-acoustic stimulation at various locations. *Applied Acoustics, 179*, 1–9. https://doi.org/10.1016/j.apacoust.2021.108074

Chapter 7
Design Mechanism Model of Wearing Comfort for Ear-Related Wearables and Its Application Examples

Abstract Wearing comfort of ear-related wearable devices is influenced not only by anatomical fit but also by usage motivation, scenario context, and exposure time. Existing ergonomic approaches often emphasize static structural matching and overlook the dynamic evolution of comfort perception during real-world use. To address this gap, a mechanism-based framework is developed to model wearing comfort loss as a time-dependent and context-sensitive process. Self-reported experience data were collected through semi-structured interviews and structured questionnaires under low physical-demand scenarios, including household activities, exercise, and sedentary work. The results demonstrate that comfort loss accumulates nonlinearly with continuous duration of use and varies across usage scenarios and device types. A Logistic exposure-response function is applied to quantify the relationship between usage motivation and wearing comfort loss, showing high fitting accuracy. In addition, a hierarchical static mechanism model is established by integrating auricular shape, external acoustic meatus structure, and auricle-meatus transitional features through weighted ergonomic variables. By coupling static exposure feedback with scenario-specific comfort loss models, a dynamic mechanism model for comfort perception is formed. Design rules, threshold references, and application guidelines are provided to support ergonomic evaluation and comfort-oriented design of ear-related wearables. Validation through application examples confirms strong consistency between model predictions and user-reported comfort perception.

Keywords Wearing comfort · Ear-related wearables · Usage motivation · Self-reported experience · Dynamic ergonomic model

The advancement of ear-related wearable technologies has introduced new ergonomic challenges that extend beyond static measurements and structural classifications. While earlier investigations have clarified the morphological features of the auricle and the external acoustic meatus, as well as the perceptual thresholds associated with deformation and transitional structures, these insights do not fully address how comfort perception evolves under real conditions of daily use. Ear-related wearables are rarely used in laboratory-controlled intervals; rather,

H. Fan et al., *Engineering Anthropometry of the External Ear for Human-Product Interaction*, https://doi.org/10.1007/978-981-92-2160-8_7

they are embedded in dynamic scenarios of work, exercise, and leisure, where user motivation, situational demands, and temporal exposure interact in complex ways.

One critical gap in current knowledge is the temporal dimension of comfort loss. Comfort perception is not a fixed attribute of a device-user fit but a process that unfolds with continuous usage. Devices that may feel tolerable at the beginning of an activity can gradually lead to noticeable discomfort after prolonged exposure. This progression is particularly relevant for tasks such as sedentary office work, extended cooking routines, or moderate-intensity exercise, where duration, repetition, and user attention converge to shape the trajectory of perceived comfort. Without explicit modelling of such dynamic processes, design solutions risk being optimized for initial fit while neglecting long-term usability.

A second necessity arises from the role of user motivation and context. Unlike protective devices that are often mandated for safety, ear-related wearables are typically adopted voluntarily and integrated into everyday routines. This means that the reasons for using these devices, whether to block distraction, to accompany exercise, or to provide entertainment, directly influence the tolerance and perception of comfort loss. For example, a user who employs in-ear devices during jogging may be more willing to accept minor discomfort than a user wearing similar devices during prolonged data analysis tasks. Current ergonomic models, however, rarely incorporate such motivational and contextual factors, resulting in assessments that overlook the subjective drivers of comfort perception.

Third, the absence of systematic self-report integration limits the reliability of existing comfort frameworks. While anthropometric data and laboratory measurements provide objective foundations, the subjective experiences of users are indispensable for capturing how comfort is negotiated in real contexts. Relying solely on mechanical or morphological parameters risks underestimating how users adapt, endure, or reject discomfort over time. A robust framework must therefore integrate structured self-report data with dynamic exposure modelling to ensure both validity and applicability.

Finally, the demand for mechanism-based design models underscores the urgency of this research. Ear-related wearables increasingly serve diverse populations, from office workers and athletes to industrial employees. Each group exhibits different usage intensities, exposure durations, and situational requirements. A generalized comfort framework must be flexible enough to accommodate this diversity while precise enough to provide actionable design thresholds. Developing such models requires moving beyond isolated measurements and toward a nested mechanism that links ergonomic factors, contextual variables, and temporal feedback.

In this regard, comfort perception is conceptualized not as a static state but as a dynamic mechanism shaped by time, motivation, and self-reported experience. Establishing this perspective is essential for translating ergonomic insights into practical guidelines, ensuring that future ear-related wearables are not only functionally effective but also sustainably comfortable across diverse contexts of use.

7.1 Wearing Comfort Loss Driven by Usage Motivation

This section continues the analysis of the usage motivation part presented in Chap. 2, which includes the semi-structured interview survey and the structured questionnaire survey. The results of the semi-structured interviews on usage motivation are analysed to design a questionnaire scale for wearing comfort loss driven by usage motivation. The structured questionnaire is then used to conduct a quantitative analysis of wearing comfort loss based on self-reported data, thereby providing quantitative relationships between self-reported experience and the mechanisms of physical comfort for outer-ear wearables that will inform the modelling of design mechanisms.

7.1.1 Ethics Approval

The studies of this chapter performed in line with the principles of the Declaration of Helsinki. Informed consent was obtained from all participants. The studies of this chapter were approved by the Ethics Committee of Northwestern Polytechnic University (No. IRB No. 201502024, No. 202002024).

7.1.2 Duration of Use and Usage Scenarios/Contexts (USC)

In order to obtain a more concentrated distribution for the mean single continuous duration per day, which is hereafter referred to as use intensity (hours per session·day, h/session·day), the interview results are reorganized with reference to half of an 8-h workday, that is, 4 h. In combination with the data collected in Chap. 2, there are 270 participants whose average daily duration of using related products is $\geq$4 h/day.

Participants' interview feedback on usage scenarios/contexts clusters into two categories, namely low physical-demand activities and high physical-demand activities, as summarized in Table 7.1. The low physical-demand scenarios/contexts include housework, exercise, and office work. The high physical-demand scenarios/contexts mainly involve physically demanding daily work in industrial sectors, such as construction operations, transportation infrastructure operations, mining operations, and manufacturing operations.

Under high physical-demand scenarios/contexts, the accumulation of bodily discomfort reported by users is not the primary driver that induces complete discomfort or fatigue. The dominant drivers are often environmental ergonomic factors, which include noise, thermal comfort of the user, moisture comfort at the body surface, body vibration during work, and indoor illuminance. Consequently, the quantitative relationship that links the time-varying load to perceived wearing comfort under high physical-demand activities, that is, wearing comfort loss as a function of exposure time and load, cannot be directly obtained. Therefore, the following analysis focuses on wearing comfort loss in low physical-demand scenarios/contexts [1–4]. Based on self-reported data obtained when participants are in a relaxed state and highly familiar with the sequence of events in the handling of the scenario, the process of wearing comfort loss is analysed.

According to Table 7.1, the event of the household USC is cooking, the events of the exercise USC are brisk walking (≤5 km/h) and jogging (≤10 km/h), and the event of the office USC is sedentary work. This result indicates that, with regard to low physical-demand activities, household, exercise, and office represent USCs in which ear-related wearable devices are more likely to appear. For the corresponding participants, their primary purposes of use are to pass the time, to relax their mood, and to avoid disturbance from others.

With respect to multiple-choice responses on USCs, the number of participants associated with each specific scenario/context activity is relatively close. With respect to single-choice responses on USCs, the number of participants engaged in sedentary work holds a clear advantage. Although not all individuals performing sedentary work require the use of ear-related wearable devices, the result implies that the function of simulating a disturbance-free environment will likely become one of the mainstream directions for the subdivision of relevant device design in the future. Based on the interview feedback from the corresponding participants, the use intensity of related devices under each USC is presented in Table 7.2. Among these, sedentary work demonstrates the highest use intensity, ranging from 2 to 4 h/

Table 7.1 Distribution of participants by scenario/context activities

Item	Low physical demand: N (percentage %)				High physical demand: N (percentage %)
	Cooking	Brisk walking	Jogging	Sedentary work	Industrial work in high physical-demand occupations
Daily usage scenarios/contexts of related products (multi choice)	108 (40.00%)	117 (43.33%)	96 (35.56%)	112 (41.49%)	110 (40.74%)
Daily most frequently used scenario/context of related products (single choice)	22 (8.12%)	23 (8.52%)	30 (11.11%)	85 (31.48%)	110 (40.74%)

Number of participants with daily usage time ≥4 h, total N = 270

Table 7.2 Time required for scenario/context activities

Duration of use		Low physical demand				High physical demand
		Cooking	Brisk walking	Jogging	Sedentary work	Industrial work in high physical-demand occupations
Experienced usage intensity	h/time·day	1–2	1–2	1–2	2–4	≥4

session*day, whereas the intensities for the other USCs remain within 1 to 2 h/ session*day.

7.1.3 Wearing Comfort Loss

A questionnaire scale for wearing comfort loss driven by usage motivation was designed with reference to the usage scenarios or contexts (USCs) and the continuous duration of use, as shown in Table E.1. According to Table 7.2, the measurement events of this questionnaire scale include cooking in the household USC, brisk walking and jogging in the exercise USC, and sedentary work in the office USC.

In terms of classification based on body-device contact type in comfort perception measurement, cup-type and cover-type devices are considered non-invasive devices, whereas band-type and plug-type devices are considered invasive devices. Accordingly, the scale sets two device items for each event, and participants fill in one or both items depending on their actual situation. The comfort perception feedback of invasive devices results from the combined influence of anthropometric dimensions and functional factors. As demonstrated in Chap. 5 of this book, the comfort perception feedback related to anthropometric dimensions constitutes the foundation of physical comfort perception for invasive devices. Therefore, within the scale, the numerical range of comfort loss for non-invasive devices is set at 0 to -5, while that for invasive devices is set at 0 to -10.

Self-reported data were collected under the condition that participants were in a relaxed state and were highly familiar with the entire process of the given scenario events. Consequently, the scale specifies the following conditions for data collection: if a participant does not have long-term experiences or familiarity that meet the following criteria in a given USC, the question can be left unanswered:

1. The device used is considered by the participant to be the most optimal product in terms of comfort.
2. The participant is in a physically and mentally relaxed state within the given scenario or context.
3. The participant is highly familiar with the entire process of the given scenario or context event.
4. Sweating is not the cause of the decline in comfort for the participant.

In addition, participants are instructed to mark the time load according to their usual use intensity. If their regular use intensity is higher than the given time load, no additional entries are required. If their regular use intensity is lower than the given time load, they should mark the specific time option that matches their actual use intensity. The number of valid questionnaires for each event is presented in Table 7.3. For non-invasive devices, the number of participants completing cooking, brisk walking, and jogging events did not exceed 274, which is the minimum required sample size. Therefore, no further analysis is conducted for these parts of the data.

Table 7.3 Questionnaire results of scenario/context activities

Usage scenarios/contexts	Number of valid questionnaires	
	Non-invasive devices	Invasive devices
Cooking	16	365
Brisk walking	46	443
Jogging	51	420
Sedentary work	306	322

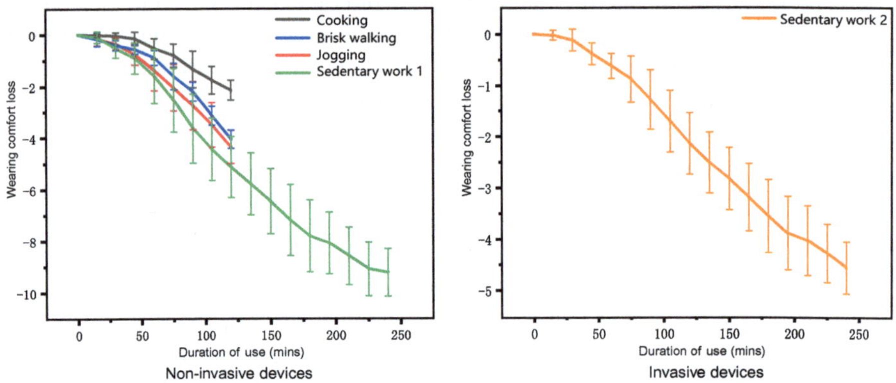

Fig. 7.1 Wearing comfort loss of motivation

As shown in Fig. 7.1, under the condition that participants were relaxed and familiar with the completion process, the cooking event demonstrated the lowest efficiency of comfort loss when using invasive devices. By the end of 120 min, the cumulative comfort loss amounted to −2.14 ± 0.38 for the cooking event, −4.05 ± 0.35 for the brisk walking event, and −4.02 ± 0.64 for the jogging event. These results are generally consistent with the common-sense comparison of physical exertion levels among the three events [3, 4].

In combination with Chap. 6, it can be observed that at 120 min the cooking event approximately corresponds to the loss of comfort that is established by the functional dimension, while the brisk walking and jogging events approximately correspond to the combined loss of functional-dimension comfort and mixed comfort, the latter being established jointly by functional and anthropometric dimensions. Without considering the specific numerical settings, the variation trends of comfort loss values during the sedentary work event are similar between invasive and non-invasive devices. The numerical results of the wearing comfort loss process driven by usage motivation are presented in Table E.2.

7.1.4 Self-Reported-Experience Model of Usage Motivation and Comfort Loss

The average variation trend of comfort loss shown in Fig. 7.1 resembles the "S" curve of the Logistic E-R function. Therefore, the Logistic E-R function is employed in this section to fit the quantitative relationship. Since the number of participants corresponding to each comfort loss value is not equal (Table E.2), the fitting is conducted using all individual participants' self-reported-experience data, rather than fitting mean values. The general form of the self-reported-experience quantitative relationship between usage motivation and comfort loss based on the Logistic E-R function is expressed as follows:

$$L_j = \frac{A_1 - A_2}{1 + \left(\frac{t}{t_0}\right)^p} + A_2 \tag{7.1}$$

where, L_jj denotes the wearing comfort loss under a given usage scenario or context, t represents the continuous duration of use, t_0 indicates the inflection point of the curve, A_1 refers to the initial value, A_2 represents the terminal value, and p is the exponent.

The Adjusted-R^2 of curve fitting ranges from 0.916 to 0.959 (Table 7.4). The results of the curve fitting are illustrated in Figs. 7.2 and 7.3, and the threshold values of wearing comfort loss driven by usage motivation, which can be conveniently referenced, are provided in Table 7.5. The models of usage-motivation-driven comfort loss corresponding to each event are expressed as follows:

$$L_{cooking} = \frac{3.48864}{1 + \left(\frac{t_1}{104.27341}\right)^{3.37877}} - 3.48351, \tag{7.2}$$

$$L_{walking} = \frac{41.42980}{1 + \left(\frac{t_1}{321.58116}\right)^{2.27726}} - 41.52763 \tag{7.3}$$

Table 7.4 Fitting for wearing comfort loss process of motivation

Event	Device type	Fitting R^2	Adjusted-R^2
Cooking	Invasive	0.91793	0.91668
Brisk walking	Invasive	0.90952	0.90923
Jogging	Invasive	0.91741	0.91654
Sedentary work 1	Invasive	0.93915	0.93895
Sedentary work 2	Non-invasive	0.95947	0.95934

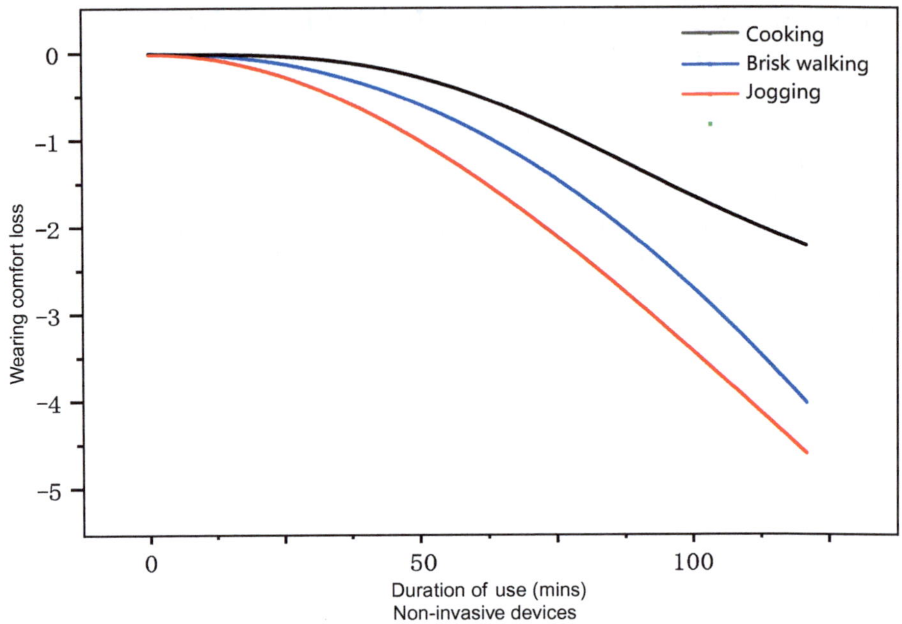

Fig. 7.2 Curve fitting for wearing comfort loss process of motivation (housework and exercise)

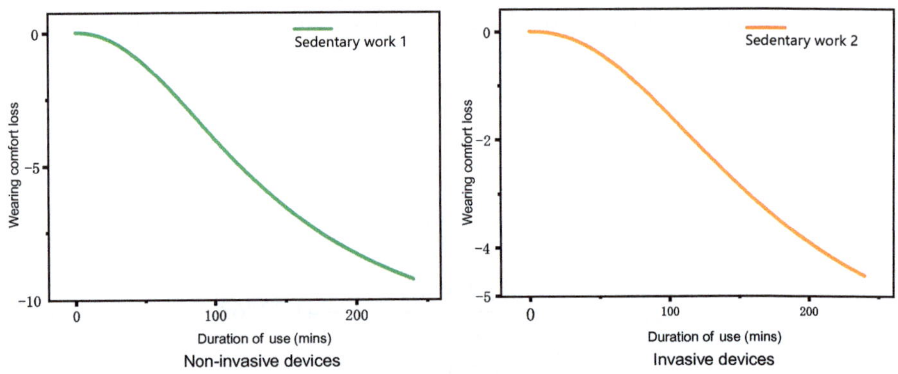

Fig. 7.3 Curve fitting for wearing comfort loss process of motivation (sedentary work)

Table 7.5 Wearing comfort loss thresholds for motivation

Duration of use (mins)	Comfort loss				
	Cooking (invasive device)	Brisk walking (invasive device)	Jogging (invasive device)	Sedentary work 1 (invasive device)	Sedentary work 2 (non-invasive device)
	0 to −10	0 to −10	0 to −10	0 to −10	0 to −5
0	0.00	0.00	0.00	0.00	0.00
15	−0.0002	−0.14	−0.11	−0.09	−0.03
30	−0.04	−0.28	−0.37	−0.43	−0.14
45	−0.19	−0.56	−0.79	−1.01	−0.33
60	−0.46	−0.98	−1.35	−1.73	−0.60
75	−0.86	−1.55	−2.01	−2.57	−0.93
90	−1.31	−2.26	−2.75	−3.44	−1.30
105	−1.76	−3.10	−3.53	−4.31	−1.70
120	−2.14	−4.06	−4.33	−5.12	−2.10
135	/	/	/	−5.88	−2.49
150	/	/	/	−6.57	−2.86
165	/	/	/	−7.17	−3.21
180	/	/	/	−7.71	−3.53
195	/	/	/	−8.18	−3.82
210	/	/	/	−8.59	−4.08
225	/	/	/	−8.95	−4.32
240	/	/	/	−9.26	−4.53

$$L_{jogging} = \frac{16.65653}{1+\left(\frac{t_1}{202.83575}\right)^{2.00685}} - 16.67939 \tag{7.4}$$

$$L_{sitting-work-1} = \frac{12.20895}{1+\left(\frac{t_2}{139.3754}\right)^{2.12178}} - 12.19604 \tag{7.5}$$

$$L_{sitting-work-2} = \frac{6.61226}{1+\left(\frac{t_2}{169.29402}\right)^{2.22354}} - 6.61273 \tag{7.6}$$

where, $L_{cooking}$ represents the wearing comfort loss during the cooking event in the household usage scenario or context when using invasive devices, $L_{walking}$ denotes the wearing comfort loss during brisk walking in the exercise usage scenario or context when using invasive devices, and $L_{jogging}$ refers to the wearing comfort loss during jogging in the exercise usage scenario or context when using invasive devices. $L_{sitting-work-1}$ indicates the wearing comfort loss during sedentary work in the office usage scenario or context when using invasive devices, while $L_{sitting-work-2}$

denotes the wearing comfort loss during sedentary work in the office usage scenario or context when using non-invasive devices. The variable ttt represents the continuous duration of use, $t_1 \in [0,120]$, $t_2 \in [0,240]$.

7.1.5 Wearing Comfort Loss, Time, and Physical Fatigue

In the absence of pain, the dynamic indicator of a state of complete loss of physical comfort is physical fatigue [5]. Under the condition that participants are relaxed and familiar with the process, the accumulation of bodily discomfort (defined in this book as wearing discomfort) during low physical-demand events is the primary factor that induces physical fatigue. According to the statistical results of the self-reported-experience data (Fig. 7.1), the amount of wearing comfort loss during cooking, brisk walking, and jogging remains within half of the global scale within 120 min. In contrast, the degree of wearing comfort loss during sedentary work almost reaches the state of complete discomfort at 480 min, which indicates a high probability of inducing physical fatigue. Based on Fig. 7.1, the efficiency of wearing comfort loss increases with the level of physical exertion. In addition to the degree of physical exertion, time plays a decisive role in the dynamic process of wearing comfort loss, as the cumulative loss gradually increases with continuous duration of use until participants establish a feedback state of complete discomfort.

Ear-related wearable devices are fitted to the external ear, which constitutes part of the head-face system. The biomechanical intensity of head movements also influences the process of wearing comfort loss [6–15]. Laboratory-based simulations were conducted for low physical-demand events and selected high physical-demand events, such as lifting, pushing and pulling, carrying, and overhead operations [3, 4] (Fig. 7.4). Head movement intensity was measured using Xsens 3D Motion Capture (Xsens Technologies B.V., the Netherlands), which integrates 17 human inertial measurement units and involved 15 participants. The measurements included angular acceleration of head movements, swing frequency, swing duration, and activity duty cycle. The results did not reveal any significant correlations among events in terms of variations in head biomechanical parameters, nor were any convertible relationships identified among the parameters.

Therefore, based on the fitting results, only the general form of the self-reported-experience quantitative relationship (Eq. (7.1)) and the specific comfort loss functions for individual events are provided, while a generalized function of comfort loss across events was not derived. Consequently, the results cannot be applied to the evaluation of wearing comfort loss in high physical-demand events. Since no generalized loss function across events was obtained, the evaluation of approximate wearing comfort loss for events outside those given may be subject to limitations. Furthermore, this section is fundamentally established on the condition that participants are relaxed and familiar with the operation process. Therefore, evaluation applications that do not meet this condition may also be limited.

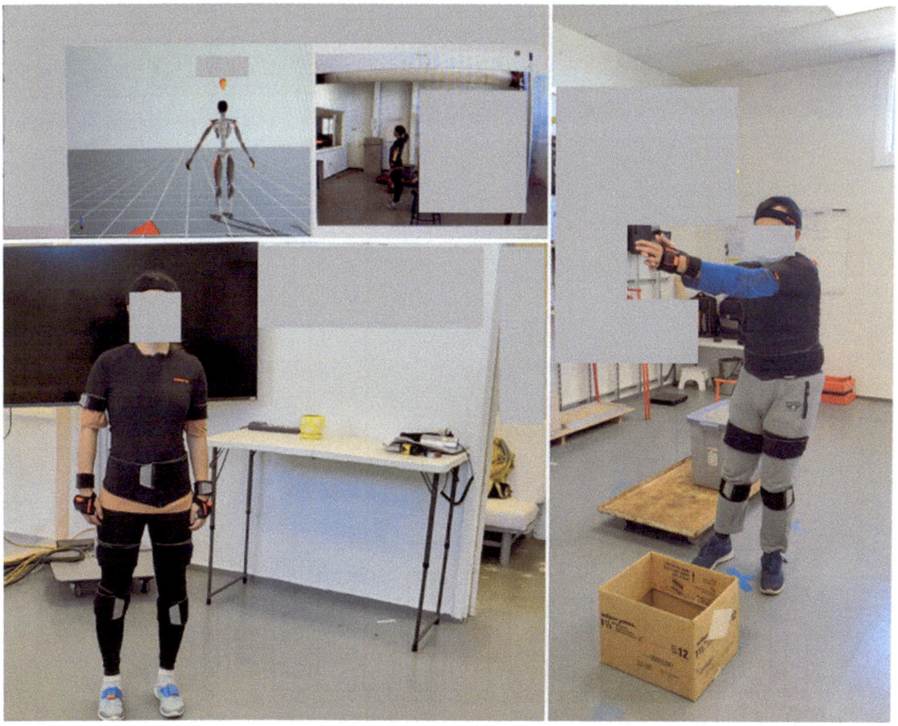

Fig. 7.4 Simulation for scenario/context activities (upper left: data interface; lower left: cooking; right: high physical demand activities)

7.2 Static Mechanism Model of Comfort Perception for Ear-Related Wearables

7.2.1 Model Construction

Based on the perspective that comfort perception feedback is controlled or co-determined by multiple factors, and in combination with Chaps. 2–6, a nested serial-structure static mechanism model of comfort perception for ear-related wearables is established, as expressed in Eq. (7.7). The complete structure of variables is illustrated in Fig. 7.5.

$$P_{Fitting} = \sum_{i=1}^{m} W_i P_i = \sum_{i=1}^{m} W_i \left(\sum_{k=1}^{n} \omega_{ik} p_{ik} \right) = \sum_{i=1}^{m} W_i \left(\sum_{k=1}^{n} \omega_{ik} \left(\sum_{h=1}^{l} u_{ik,h} g_{ik,h} \right) \right) \quad (7.7)$$

where, l denotes the number of variable items contained in the k-th subfactor of the i-th ergonomic factor; n denotes the number of subfactors in the i-th ergonomic design factor; and m denotes the total number of ergonomic design factors; h refers to the h-th variable item within the k-th subfactor; p represents the user's perception of wearing

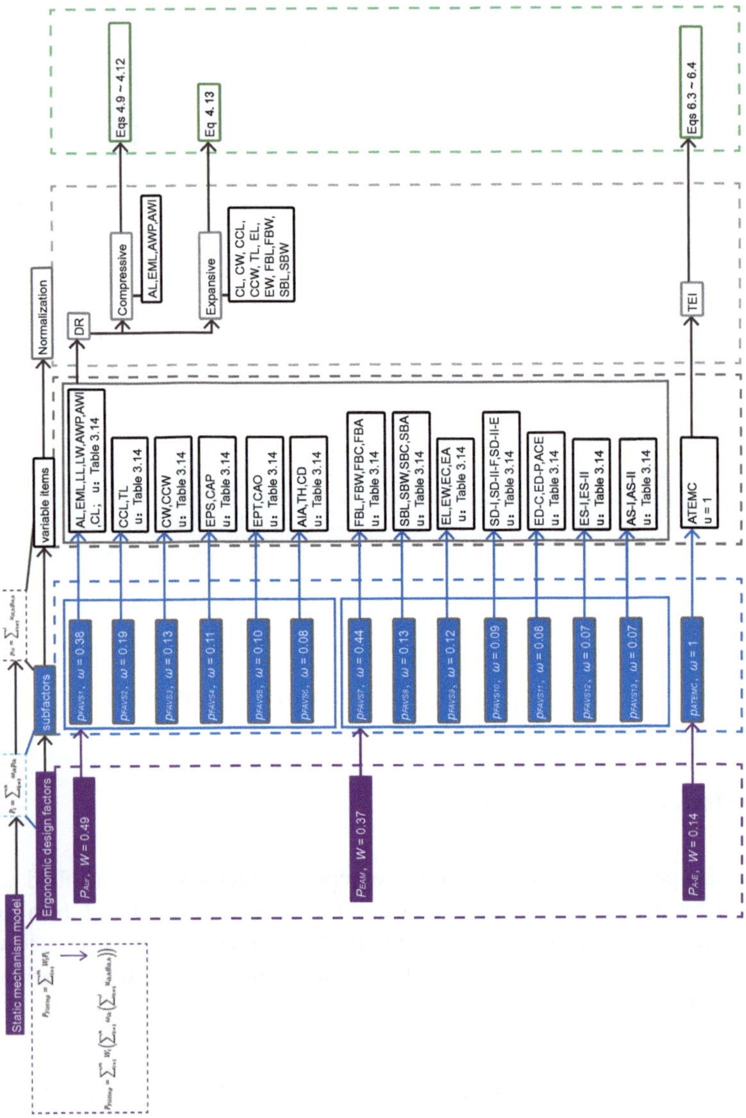

Fig. 7.5 Static exposure models for ear-related wearing comfort perception

comfort loss with respect to changes in the subfactor, while g represents the user's perception of wearing comfort score with respect to changes in the variable items of the subfactor; the weight W corresponds to the ergonomic design factor, the weight ω corresponds to the subfactor, and the weight u corresponds to the variable item.

The static mechanism model of comfort perception for ear-related wearables is organized in a hierarchical structure consisting of three layers, namely the ergonomic design factor layer, the secondary model layer, and the tertiary model layer. The first layer of the static mechanism model includes three ergonomic design factors, which are the auricular shape-structure factor, the external acoustic meatus shape-structure factor, and the auricle-meatus transitional structure factor. The weights (W) of each ergonomic design factor are specified in Eq. (2.13).

The second layer is the group of secondary models, which consists of three models. Each model is constructed using an ergonomic design factor together with its corresponding FASVs. The FASVs of the auricular shape-structure factor include auricular contour (FASV 1), tragus-concha length (FASV 2), conchal width (FASV 3), superior auricular protrusion (FASV 4), posterior auricular protrusion (FASV 5), and age-independent auricular items (FASV 6), with their weights (ω) detailed in Eq. (3.3). The FASVs of the external acoustic meatus shape-structure factor include first bending of the external acoustic meatus (FASV 7), second bending of the external acoustic meatus (FASV 8), external acoustic meatus entrance (FASV 9), meatal segment depth (FASV 10), external acoustic meatus entrance position (FASV 11), meatal segment height angle (FASV 12), and meatal segment azimuth angle (FASV 13), with their weights (ω) specified in Eq. (3.4). The auricle-meatus transitional structure factor contains only ATEMC (angle between tragus, entrance, and medial concha, FASV 14), whose weight (ω) is 1.

The third layer is the group of tertiary models, which consists of 14 models in total. In the specific studies presented in each chapter, the FASVs and their variable items are organized through dimensionality reduction or comparative analysis of variability. Auricular contour (FASV 1) contains seven variable items, tragus-concha length (FASV 2) contains two variable items, conchal width (FASV 3) contains two variable items, superior auricular protrusion (FASV 4) contains two variable items, posterior auricular protrusion (FASV 5) contains two variable items, and age-independent auricular items (FASV 6) contain three variable items. Their weights (u) are provided in Table 3.14. First bending of the external acoustic meatus (FASV 7) contains four variable items, second bending of the external acoustic meatus (FASV 8) contains four variable items, external acoustic meatus entrance (FASV 9) contains four variable items, meatal segment depth (FASV 10) contains three variable items, external acoustic meatus entrance position (FASV 11) contains three variable items, meatal segment height angle (FASV 12) contains two variable items, and meatal segment azimuth angle (FASV 13) contains two variable items. Their weights (u) are provided in Table 3.14. ATEMC (FASV 14) contains one variable item, and its weight (u) is 1. The hierarchical structure of the static mechanism model is illustrated in Fig. 7.5.

The fourth layer is the group of quaternary models, which are calculated separately according to the variations of individual variables. In this book, the variation degree of each variable is standardized, and the standardized variables are then used to establish quantitative relationships between variable changes and wearing comfort scores in terms of perception feedback.

Specifically, the deformation ratio (DR) is used to standardize the variations of the auricular shape-structure factor and the external acoustic meatus shape-structure factor, while the tragus expansion index (TEI) is used to standardize the variations of the auricle-meatus transitional structure factor. According to the research findings on the auricular and external acoustic meatus shape-structure factors, AL, EML, AWP, and AWI are defined as compressive deformation variables, whereas CL, CW, CCL, CCW, TL, EL, EW, FBL, FBW, SBL, and SBW are defined as expansive deformation variables. The remaining variables of the auricular and external acoustic meatus shape-structure factors are quantitatively defined as functional variables, which are considered free of discomfort provided that their design values do not induce restrictions on head movements or cause the device to fall off the ear, or are computed theoretically by using a 5% variation.

The general form of the perception feedback model for wearing comfort loss with respect to changes in compressive deformation ratio is expressed as follows:

$$g_{DR-ext} = y_0 + Ae^{-\frac{DR_{ext}}{t}} \tag{7.8}$$

where, y_0 denotes the offset, A represents the amplitude, and t refers to the period intercept; g_{DR-ext} denotes the user's perception of wearing comfort score with respect to changes in the compressive deformation ratio, and DR_{ext} denotes the deformation ratio produced by compression of the relevant variable during fitting due to structural constraints.

The general form of the perception feedback model for wearing comfort loss with respect to changes in expansive deformation ratio is expressed as follows:

$$g_{DR-expa} = a\left(1 - e^{-bDR_{expa}}\right) \tag{7.9}$$

where, a denotes the coefficient, b denotes the coefficient, $g_{DR-expa}$ denotes the user's perception of wearing comfort score with respect to changes in the expansive deformation ratio, and DR_{expa} denotes the deformation ratio generated by expansive structural change of the relevant variable during product fitting.

The general form of the perception feedback model for wearing comfort loss with respect to changes in the expansive degree of the auricle-meatus transitional structure is expressed as follows:

$$g_{TEI} = y_0 + Ae^{-\frac{(TEI - x_c)^2}{2w^2}} \tag{7.10}$$

where, y_0 denotes the offset, x_c denotes the central point, and w denotes the width with $w = \frac{FWHM}{2\sqrt{\ln 4}}$; A represents the amplitude, g_{TEI} denotes the user's perception of wearing comfort score with respect to changes in the expansive degree of the auricle-meatus transitional structure, and TEI denotes the tragus expansion index.

This book provides threshold lookup tables for rapid estimation of the above models, as well as reference tables of design values based on the mean statistical results of the participants. The threshold lookup table for rapid estimation of *DRext* and *DRexps* is presented in Table 4.3. The design values of other functional variables can be referenced using the data of female participants as the baseline, and theoretical computation is performed using a 5% variation, as shown in Table 4.4. The threshold lookup tables for rapid estimation of TEI are presented in Tables 6.1 and 6.2, and the reference table of design values for ATEMC is provided in Table 6.3.

7.2.2 *Comfort Design and Rules for Application*

Rule 1

According to the design requirements of actual products, the corresponding ergonomic design factors, subfactors, and their variable items should be selected. Based on subfactors and their variable items, a subfactor-product matching network framework is provided to support rapid selection of variables for the application of comfort schemes (Fig. 7.6). On this basis, existing products can be broadly categorized into six types:

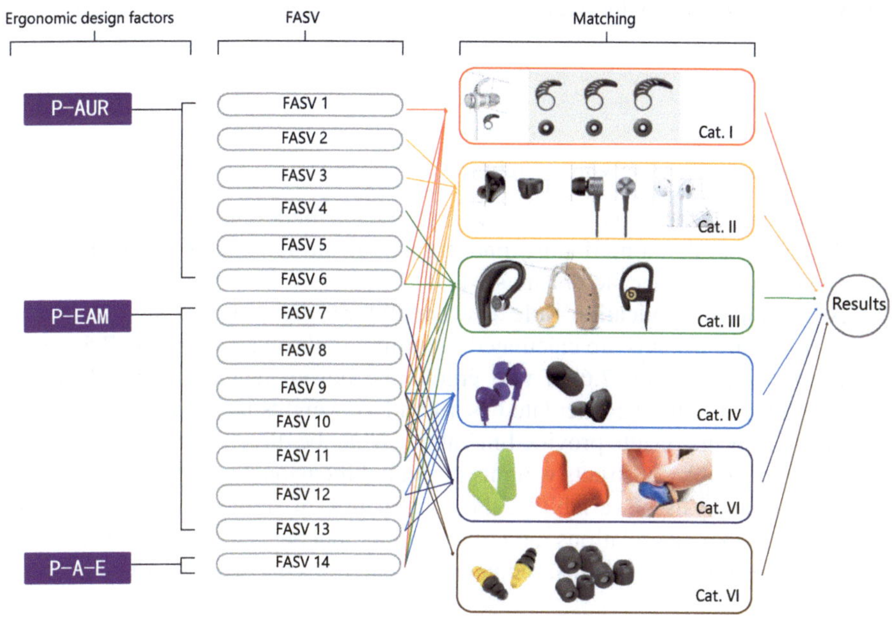

Fig. 7.6 Subfactor-product matching networks

Cat. I: Plug-type products with wing-shaped fixers. The wing-shaped fixer causes an expansive deformation along the conchal length direction. This type mainly involves variable items from subfactors 1, 6, 9, 11, and 14.

Cat. II: Non-angled plug-type products. The loudspeaker and the main body of the product are designed without an angle. This type mainly involves variable items from subfactors 2, 3, 6, 9–11, and 14.

Cat. III: Band-type products. This type mainly involves variable items from subfactors 4–6, 9–11, and 14.

Cat. IV: Angled invasive products. The loudspeaker and the main body of the product are designed with an angle, and the fixation does not rely on the concha. This type mainly involves variable items from subfactors 9–14.

Cat. V: Soundproof earplugs and in-ear enhancement devices. This type mainly involves variable items from subfactors 7–10 and 12–13.

Cat. VI: Waterproof earplugs and anti-fall earplugs. This type mainly involves variable items from subfactors 9–11.

Rule 2

According to the comfort thresholds and exposure limits provided by this book, if a product is designed to cover all ergonomic design factors, the highest recommended wearing comfort score calculated by the static mechanism model is 0.35 ($0 \times 0.49 + 0 \times 0.37 + 2.5 \times 0.14$), which represents a mixed wearing comfort feedback result. The lowest wearing comfort score that can be calculated is −4.93 ($-5 \times 0.49 + -5 \times 0.37 + -4.5 \times 0.14$), which represents the wearing comfort feedback result.

Rule 3

The specific rules for applying weights are as follows:

(a) According to the definition of the single-factor model in Eq. (2.2), in scheme design or evaluation, if only one ergonomic design factor is involved, the weights of the other factors are set to 0.
(b) According to Application Rule 2, when all three ergonomic design factors are covered, the weights are calculated according to those given in Fig. 7.5.
(c) In reference to Fig. 7.6, if two ergonomic design factors are involved, the weights can either be calculated as secondary weights based on those given in Fig. 7.5, or the results provided in Table 7.6 can be directly adopted.
(d) Since this book constructs weights through dimensionality reduction and comparative analysis of variability, the construction of subfactors is based on the inter-variable associations and inter-group independence. Therefore, when applying subfactor weights, secondary calculations may be performed in accor-

Table 7.6 Weight distribution (*W*) of ergonomic factors under different conditions

Condition	Weight distribution (*W*)		
	Auricular shape-structure factor	External acoustic meatus (EAM) shape-structure factor	Auricle-EAM transitional structure factor
Single-factor case 1	1.00	/	/
Single-factor case 2	/	1.00	/
Single-factor case 3	/	/	1.00
Two-factor case 1	0.57	0.43	/
Two-factor case 2	/	0.63	0.37
Two-factor case 3	0.78	/	0.22
Three-factor case	0.49	0.37	0.14

dance with Rules 3-a and 3-c, namely by recalculating weights within the ergonomic design factors that are practically involved. When calculating the weights of variable items, all variable items within the subfactor must participate in the calculation. The comfort score of variable items that do not change is set to 0, while the variable items that change must be calculated using the weights provided in this book.

Rule 4

In combination with the user's expected usage intensity, the comfort score should be determined accordingly. The recommended range of overall comfort scores in the static mechanism model of comfort perception for ear-related wearables is [−0.5,0]. For invasive devices, the recommended range is [0,2.5].

Rule 5

Based on the determined comfort score, the perception feedback model of wearing comfort with respect to variable-item changes can be applied to calculate the specific variation of each variable. Alternatively, the provided threshold lookup tables can be used for direct reference.

Rule 6

After determining or calculating the threshold of variation, the design value scheme should be formulated according to the specific target users. For universal designs intended for both males and females, it is recommended to adopt as the baseline the sex group that shows a significant disadvantage in terms of comfort perception. If no significant sex difference is observed, the overall mean may be used for calculation. For universal designs targeting specific age segments or percentile ranges, the corresponding results provided in the Appendix can be used for further calculation.

Rule 7

When conducting evaluation studies of design schemes, it is necessary to first identify the specific target population and then calculate the degree of variable-item changes induced by the design parameters.

Rule 8

When conducting evaluation studies of design schemes, according to the definition in Eq. (2.2), if the design value of any item exceeds the exposure limit, the user will be unable to establish wearing comfort perception, and the corresponding product parameter becomes invalid.

7.2.3 Recommendations for Comfort Design Application

Comfort Design

In the formulation or evaluation of design schemes, the mean thresholds and exposure limits of wearing comfort scores for variable items provided in this book can serve as a convenient and rapid reference for design thresholds. At the same time, the final design scheme should be determined by considering the characteristics of the specific target population, such as age group, sex, and percentile range. For design evaluations aimed at individuals or small user groups, it is recommended to incorporate the personal data of the users and conduct evaluations in combination with the variation threshold ranges provided in this book.

Sex Baseline

For auricular shape-structure variables, after removing the influence of height and weight, significant sex differences remain in AL, AWP, AWI, EPS, EPT, CAP, EL, EC, EA, FBL, FBC, FBA, SBL, SD-II-F, ED-C, ED-P, AS-I, and AS-II. Therefore, the design values of these variables should use females as the baseline, while other variables may be calculated using overall means. After eliminating the influence of height and weight, significant sex differences still exist for AL, AWP, AWI, EL, FBL, and SBL. Regarding deformation ratios, in universal designs for both sexes, it is recommended that the design values of AL, AWP, AWI, EL, FBL, and SBL be calculated using the female data.

For auricle-EAM transitional structures, after eliminating the influence of height and weight, the sex difference in ATEMC is not significant; thus, the design baseline for ATEMC should use the overall mean. Based on the exposure-response model, the female TEI range from TEI(2.5) to TEI(0) (11.48%–19.54%, mixed comfort threshold) is more suitable for universal design applications. If the design is exclusively for male users, the applicable range should be TEI(3.25) to TEI(0) (4.11%–20.40%, mixed comfort threshold). For both-sex universal designs and for male-only designs, TEI values lower than 11.23% (male TEI(2.5)) are not recommended. TEI values greater than 20.20% (male TEI(−0.5)) may cause discomfort in both sexes and therefore should not be adopted for universal design applications.

Auricle-EAM Transitional Structure Expansion

According to the results of the exposure-response model of TEA/TEI, it is recommended that for invasive ear-related wearables, particularly those that are intended for sports or outdoor activities, comfort scores should be adopted within the descending zone of the tragus expansion index. The comfort score range recommended in this book is [0,2.5]. In addition, auxiliary components such as ear bands may be employed to reduce the potential risk of device detachment.

Interface Between Ear Cups or Ear Pads and Human Skin

When the external auricle and periauricular skin are uniformly loaded and the rate of change is the same, the efficiency of discomfort increases for the periauricular skin, tragus skin, and conchal skin is lower than that of other regions. Therefore, the interface between ear cups or ear pads and human skin should be designed to focus on these regions. For large ear cups, the contact areas or stabilizing reinforcement structures are recommended at the periauricular temporal bone skin, mastoid skin, tragus skin, and conchal skin. For medium-sized ear cups or pads, the contact areas or reinforcement structures should be designed at the concha and the superior rim of the helix. For small ear cups, the contact area or reinforcement structures should be designed only at the conchal skin.

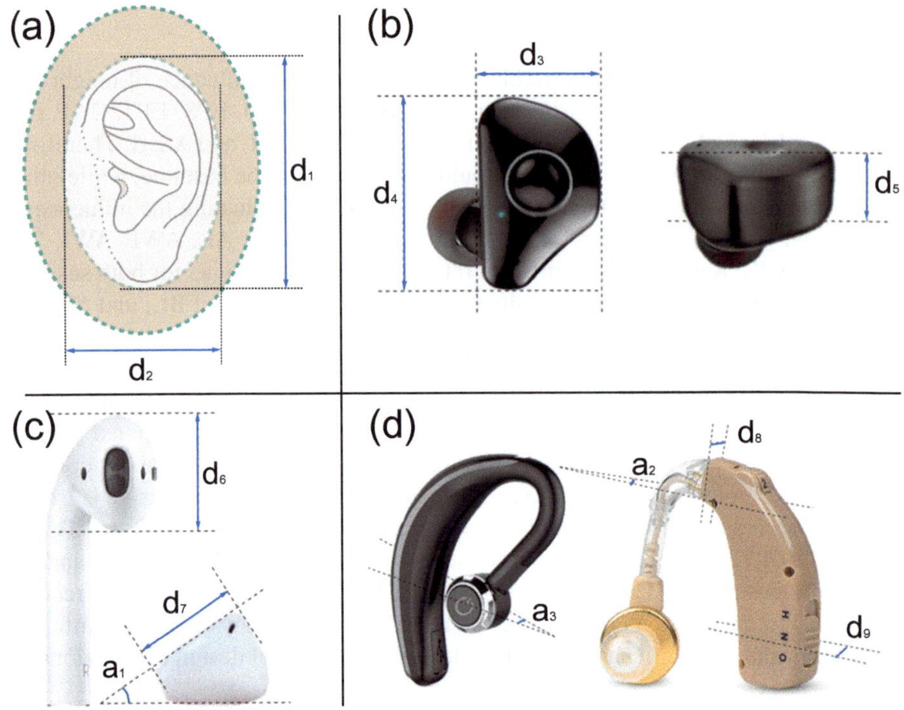

Fig. 7.7 Key measurement variables for earmuffs, earbuds and ear-bands: (**a**) earmuffs; (**b**) earbuds; (**c**) earbuds; (**d**) ear-bands

In accordance with the Chap. 4, the optimal design scheme for ear cups is recommended to involve contact only with the periauricular skin (Fig. 7.7a, in which d_1 and d_2 are both greater than the AL and AWP values of the target users). For ear pads or medium-sized ear cups, the optimal design scheme is recommended to involve contact only with the conchal skin.

Plug-Type and Band-Type Products

As shown in Fig. 7.7b, plug-type products use the entire concha as the fitting region. According to the results of Fan et al. [16, 17], the optimal design value of d_3 should be based on the user's TL value, with the specific range defined as (TL, CCL]. The optimal design value of d_4 should be based on the user's CCW value, with the specific range defined as (CCW, CW]. The parameter d_5 corresponds to the participant's CD value, and before evaluation, it is necessary to verify whether the design value restricts the user's head movement or to calculate the theoretical design value using a 5% variation.

As shown in Fig. 7.7c, another type of plug product adopts a ring-free design without attached soft rings, primarily using the auricle-EAM transitional structure as the fitting region. Therefore, it is recommended that the design range of a_1 be determined according to TEI(2.5) to TEI(0). Similar to d_4, the optimal design value of d_6 should also be based on the user's CCW value, with the specific range defined as (CCW, CW]. To ensure sufficient contact between the device and the auricle-EAM transitional structure, d_7 should be smaller than the target user's ED-P value. In evaluating the product shown in Fig. 7.7d, the design values of auricular superior protrusion (a_2 and d_9) and auricular posterior protrusion (a_3 and d_9) are required not to restrict the user's head movement, or they should be calculated theoretically using a 5% variation. When the specific levels of ED-C, ED-P, and ACE of the target user are known, trigonometric functions can be used to calculate the specific design position or coordinates of the earphone or loudspeaker in the relevant product.

Earplugs

Earplugs generally need to fit to the second segment of the external acoustic meatus (typically for standard sound insulation requirements); depending on specific design requirements, they may fit to the second bending position (for deep sound insulation requirements). In designing earplugs (Fig. 7.8), the material should, after deformation during fitting, maintain the inherent levels of ED-C, ED-P, ACE, AS-I, ES-I, AS-II, and ES-II of the user as much as possible. Earplugs are typically made of soft and malleable materials; therefore, the undeformed design value of d_{10} should be greater than the user's EL value. For ease of removal from the external acoustic meatus, C_1 and S_1 should be greater than the user's EC and EA values, respectively. For general hearing protection design requirements, the design length of d_{11} should be greater than SD-I and smaller than SD-II-E. For deep hearing protection design requirements, the design length of d_{11} should be greater than SD-II-E and smaller than the sum of SD-I and SD-II-F.

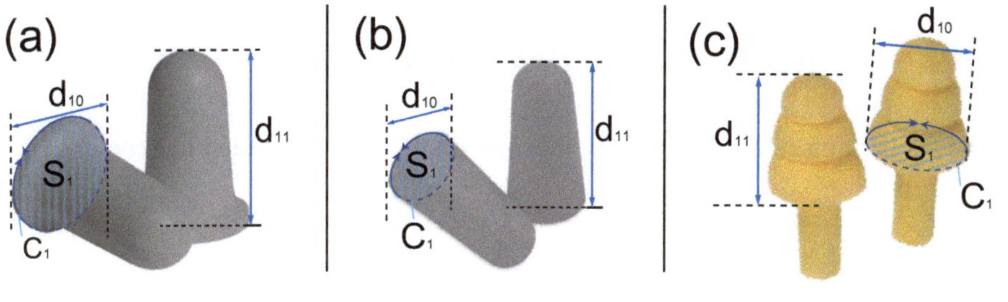

Fig. 7.8 Key measurement variables for earplugs: (**a**) earplugs (with a base-layer); (**b**) earplugs (uncorded); (**c**) waterproof earplugs

7.3 Dynamic Mechanism Model of Comfort Perception for Ear-Related Wearables

7.3.1 Model Construction

The general form of the dynamic mechanism model is expressed as follows:

$$P_{Motive} = P_{Fitting} + L_j = \sum_{i=1}^{m} W_i \left(\sum_{k=1}^{n} \omega_{ik} \left(\sum_{h=1}^{l} u_{ik,h} g_{ik,h} \right) \right) + \frac{A_1 - A_2}{1 + \left(\frac{t}{t_0} \right)^p} + A_2 \quad (7.11)$$

The calculation result(s) of the dynamic mechanism model(s) is the sum of the static mechanism model(s) and the comfort-loss model(s). The structure of the comfort-loss model consists of four components: usage scenarios or contexts, scenario events, continuous duration of use, and self-reported-experience quantitative relationships. As no parameter-based conversion relationships of scenario events are derived, the comfort-loss model could be formulated as a set of five quantitative relationship models based on self-reported experiences.

The usage-scenario/context component includes three specific scenarios: household, exercise, and office. The specific event of the household scenario is cooking with invasive devices, with a continuous duration of 0–120 min, and the corresponding quantitative relationship model is expressed in Eq. (7.2). The specific events of the exercise scenario are brisk walking and jogging with invasive devices, each with a continuous duration of 0–120 min, and the corresponding quantitative relationship models are expressed in Eqs. (7.3) and (7.4), respectively. The specific event of the office scenario is sedentary work, performed with invasive devices and with non-invasive devices, each with a continuous duration of 0–240 min, and the corresponding quantitative relationship models are expressed in Eqs. (7.5) and (7.6), respectively. Table 7.5 provides the threshold lookup table for rapid estimation of the above models.

7.3.2 Comfort Design and Rules for Application

The application rules for the static mechanism model are detailed in Sect. 7.2.

Rule 9

When applying the dynamic mechanism model, it is first necessary to calculate or to measure the static exposure feedback of the target user. For scheme design and evaluation, the process begins by calculating the static exposure feedback based on the difference between the design values and the parameters of the target user. The comfort-loss model is then employed to calculate the process of comfort loss corresponding to the expected usage intensity, after which the calculation results are used to establish or to optimize the design usage intensity of the scheme. For evaluations targeting individual users, the relevant variable items of the user must first be measured. The static mechanism model is then used to calculate the comfort score of the test product under static fitting. This is subsequently combined with the user's habitual usage intensity to calculate the process of comfort loss, and finally, recommendations are provided either for adjusting the usage intensity or for replacing the product.

Rule 10

The comfort-loss model provides quantitative relationship models corresponding to specific events. If the scheme design and evaluation concern a given event, the corresponding model can be directly applied. If the scheme design and evaluation involve events that are similar in terms of physical exertion to the given events, and relevant evidence can be found in published sources (for example, the physical exertion of indoor cleaning work is close to that of brisk walking [190,219]), then the model can be approximately applied.

7.3.3 Examples of Wearing Comfort Perception Feedback Assessment

By conducting comfort perception feedback evaluations of two different products, the practical application of the ear-related wearable comfort perception model is further demonstrated. At the same time, the usability and validity of the model are examined through comparative analysis of the differences between the model-based calculation results and the participants' reported comfort perception feedback.

Example 1: Cooking with Invasive Earphones for 60 Min

In this example, a female participant was provided with the product shown in Fig. 7.7c. The assessment side was the left ear, and the usage motivation under assessment was the cooking event within 60 min (Fig. 7.9a).

The participant was 30 years old, with a daily cooking time of approximately one and a half hours. The cooking usually included meals for the evening as well as breakfast and lunch for the following day. During these household USC events, the participant habitually listened to music at an extremely low volume, a practice that had been maintained for 3 years. The specific reason for this behaviour was to

Fig. 7.9 Assessment of invasive devices: (**a**) cooking; (**b**) sedentary work

relieve psychological stress accumulated during daytime work through the slow rhythm of the cooking process, while the extremely low volume setting was intended to avoid a "deafening" sensation when unexpectedly receiving phone calls from family members, friends, or delivery personnel (including food and beverage delivery).

This example was conducted under the experimental condition of the participant's daily behavioural habits. The volume was set at the lowest level, and the playlist consisted of the participant's regular light-music playlist with a duration of nearly 2 h, with tracks played in random order. This example recorded only the first 60 min of the event process. The recording method was unsupervised video capture, and the starting point of the recording was set at the participant's first step in the daily process, namely opening the refrigerator to retrieve vegetables.

Step 1: Product Parameters and Measurement Variables

The required product variables were d6, d7, and a1 in Fig. 7.7c, corresponding to the participant's TL (d6), CCL (d6), ED-P (d7), and ATEMC (a1). These variables involved the tragus-concha length of the auricular shape-structure factor, the entrance position of the external acoustic meatus shape-structure factor, and the external-expansion of the auricle-EAM transitional structure factor. The measurement results of the product assessment parameters and participant variables are presented in Table 7.7.

Step 2: Static Exposure Feedback

Before the assessment, the participant was first asked to test whether the provided product would restrict head movement. After performing actions such as head shaking and jumping, no detachment of the test product occurred. Therefore, the variations in TL and ED-P of the participant were marked as non-restrictive for head movement. After concha casting and three-dimensional scanning, the changes in the

Table 7.7 Comparisons for variables (Example 1)

Item	Variable product		Participant		Deformation (%)	
1. (mm)	d_6	18.50	TL	16.23	Inhibit head movement	No
2. (mm)	d_6	18.50	CCL	17.08	DR_{expa}	8.31
3. (mm)	d_7	12.02	ED-P	10.39	Inhibit head movement	No
4. (°)	a_1	30.32	ATEMC	26.04	TEI	16.43

participant's CCL and ATEMC were measured as 8.31% (expansive deformation) and 16.43% (TEI), respectively (Table 7.7). Within the tragus-concha length subfactor, the weights of TL and CCL are both 0.50 (Table 3.14). According to the application rules, the static exposure assessment model for the participant using the test product was therefore established as follows:

$$\begin{aligned} P_{Fitting} &= 0.49p_{Aur} + 0.37p_{EAM} + 0.14p_{A-E} = 0.49g_{DR-expa} + 0.37\times 0 + 0.14g_{TEI} \\ &= 0.49\times 0.50g_{CCL} + 0.49\times 0 + 0.37\times 0 + 0.14g_{TEI} \end{aligned} \tag{7.12}$$

By applying the calculation formulas of g_{CCL} and g_{TEI}, the participant's static exposure feedback of wearing comfort for ear-related wearables was approximately −0.5, which corresponds to ambiguous discomfort.

Step 3: Dynamic Exposure Feedback

The dynamic exposure assessment model for the participant using the test product was established as follows:

$$P_{Motive} = P_{Fitting} + L_{cooking} = -0.5 + \frac{3.48864}{1+\left(\frac{t_1}{104.27341}\right)^{3.37877}} - 3.48351 \tag{7.13}$$

According to the assessment model, the participant's dynamic exposure feedback of wearing comfort for ear-related wearables at 0, 15, 30, 45, and 60 min was approximately −0.5, −0.5, −0.5, −0.7, and −1, respectively.

Step 4: Output

The comfort score scale ranges from +5 to −5: extreme comfort (+5), very high comfort (+4.5 and +4), high comfort (+3.5 and +3), moderate comfort (+2.5 and +2), little comfort (+1.5 and +1), just noticeable comfort (+0.5), normal/no discomfort (0), just noticeable discomfort (−0.5), little discomfort (−1 and −1.5), moderate discomfort (−2 and −2.5), high discomfort (−3 and −3.5), very high discomfort (−4 and −4.5), and extreme discomfort (−5).

Before the recording, the participant rated the test product with a comfort score of −0.5. During the recording, the participant provided a comfort score every 15 min, with increments marked at integer multiples of 0.5. The participant's dynamic exposure feedback scores of wearing comfort at 0, 15, 30, 45, and 60 min were −0.5, −0.5, −0.5, −0.5, and −1, respectively.

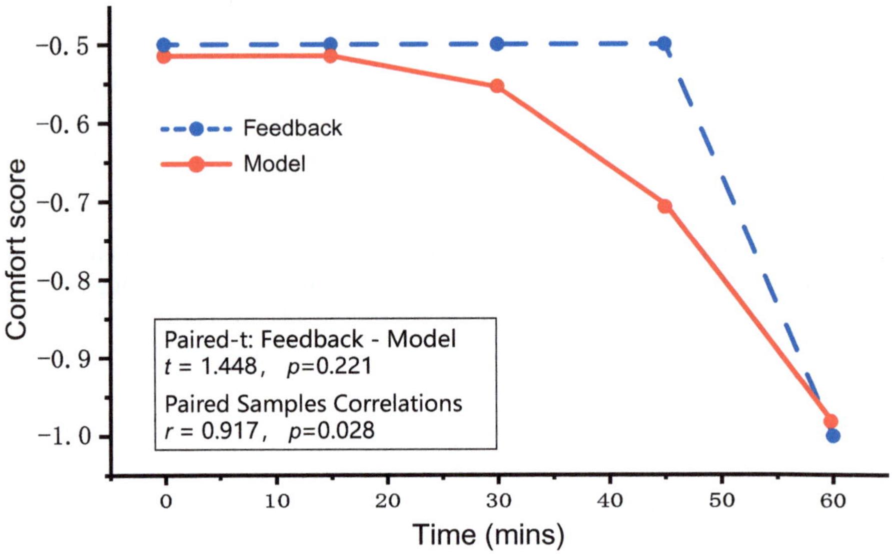

Fig. 7.10 Comparison of results for Example 1

The comparison between the calculated results and the participant's feedback results, as well as the paired t-test outcomes, are shown in Fig. 7.10.

Step 5: Discussion

As shown in Fig. 7.11, the overall variation trends of the calculated results and the participant's feedback results are similar, with both entering the state of little discomfort at the sixtieth minute. Because the participant's ratings were marked in increments of 0.5, the calculated results appear relatively "smoother." In other words, for the participant, in the absence of finer gradations for marking, the comfort loss at the forty-fifth minute did not induce a clearly perceptible feedback of just noticeable discomfort. According to the paired t-test, the two sets of results are highly correlated ($r = 0.917$, $p < 0.05$) and do not exhibit a significant difference ($p = 0.22$). Therefore, the model assessment results can be considered both usable and valid.

Example 2: Sedentary Work for 120 Min

This example assessed the dynamic exposure feedback of comfort for a male participant using the optimal-angle ATEMC tool, designed as a single-factor exposure assessment. The participant was recruited from Chap. 4, and the test product was the ATEMC measuring tool previously marked by the participant as providing optimal

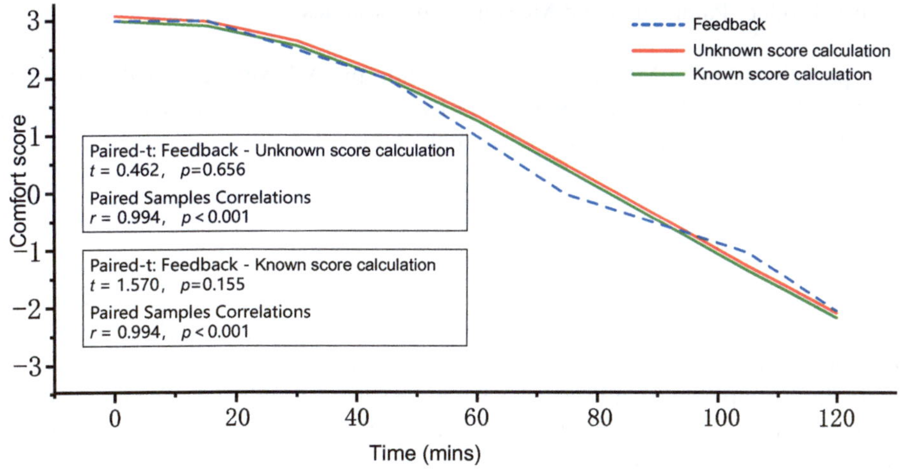

Fig. 7.11 Comparison of results for Example 2

comfort. The assessment side was the right ear, and the usage motivation was sedentary work for 120 min (Fig. 7.9b). The sedentary work task consisted of processing and analysing experimental data.

For the purpose of this example, the participant's optimal-angle ATEMC tool was manufactured. During the recording, the participant was not informed of any dimensional information regarding the test tool. The participant was 28 years old and a researcher at a scientific institution. Excluding short breaks for restroom use, his daily continuous sedentary working time was approximately 3 h. His daily work involved processing and analysing experimental data, including data import, preliminary verification, handling of missing values, label checking, synchronization testing, noise filtering, data analysis, result export, report writing, and preparation of presentation materials. He had been engaged in such work for 4 years.

During sedentary work, the participant was accustomed to using invasive devices, sometimes with music playback and sometimes without. The overall reason was to signal to surrounding colleagues that he did not wish to be disturbed. He generally did not remove the device when going to the restroom, mainly because he was often still engaged in thinking about data analysis tasks and did not want to be disturbed. The participant did not use ear cups or ear pads because, during breaks for discussions with colleagues, he found that removing and reinserting invasive devices was more convenient.

This example was conducted under the participant's daily behavioural habits as experimental conditions. The example recorded only the first 120 min of the sedentary work event. The recording method was unsupervised video capture, and the starting point of the recording was set at the first step of the participant's daily process, namely importing data.

Step 1: Product Parameters and Measurement Variables

The participant's ATEMCear was 32.11°, while the ATEMCwear of the optimal-angle tool was 35°. The corresponding TEI value was 5.71%, and the comfort score was marked as 3, which corresponds to high comfort.

Step 2: Static Exposure Feedback

Before the assessment, the participant was first asked to test whether the provided product would restrict head movement. After performing actions such as head shaking and jumping, no detachment of the test tool was observed.

In this example, two static assessment models were established. The first was the static exposure assessment model under the condition that the participant's rating of the measuring tool was known:

$$P_{Fitting} = p_{A-E} = g_{TEI} = 3 \tag{7.14}$$

The second was the static exposure assessment model under the assumption that the participant's rating of the measuring tool was unknown. Since the participant was male, the calculation of g_{TEI} was based on the male reference values. The specific form of the model is as follows:

$$P_{Fitting} = p_{A-E} = g_{TEI,male} = -4.75273 + 7.59432e^{-\frac{(TEI-7.83617)^2}{146.11225}} \tag{7.15}$$

Step 3: Dynamic Exposure Feedback

The calculation result obtained using Eq. (7.16) was 3.08, which was very close to the participant's originally marked comfort score of 3.

$$P_{Motive} = P_{Fitting} + L_{sitting-work-1} = P_{Fitting} + \frac{12.20895}{1+\left(\frac{t_2}{139.3754}\right)^{2.12178}} - 12.19604 \tag{7.16}$$

According to the assessment model, under the condition that the participant's rating was known, the dynamic exposure feedback of wearing comfort for ear-related wearables at 0, 15, 30, 45, 60, 75, 90, 105, and 120 min was 3.00, 2.91, 2.57, 1.99, 1.27, 0.43, –0.44, –1.31, and –2.12, respectively. Under the condition that the participant's rating was unknown, the corresponding dynamic exposure feedback values at 0, 15, 30, 45, 60, 75, 90, 105, and 120 min were 3.08, 2.99, 2.65, 2.07, 1.35, 0.51, –0.36, –1.23, and –2.04, respectively.

Step 4: Output

The comfort score scale ranges from +5 to −5: extreme comfort (+5), very high comfort (+4.5 and +4), high comfort (+3.5 and +3), moderate comfort (+2.5 and +2), little comfort (+1.5 and +1), just noticeable comfort (+0.5), normal/no discomfort (0), just noticeable discomfort (−0.5), little discomfort (−1 and −1.5), moderate discomfort (−2 and −2.5), high discomfort (−3 and −3.5), very high discomfort (−4 and −4.5), and extreme discomfort (−5). Before the recording, the participant marked the test tool with a comfort score of 3. During the recording, the participant reported comfort scores every 15 min, with increments marked at integer multiples of 0.5. The participant's dynamic exposure feedback scores of wearing comfort at 0, 15, 30, 45, 60, 75, 90, 105, and 120 min were 3, 3, 2.5, 2, 1, 0, −0.5, −1, and −2, respectively. The comparison between the calculated results and the participant's feedback results is shown in Fig. 7.11.

Step 5: Discussion

The calculated results and the participant's feedback results exhibited similar overall variation trends, with both reaching the state of moderate discomfort at the 120th minute. Because the participant's ratings were marked in increments of 0.5, the feedback was less "smooth" than the calculated results on both sides of the no-discomfort state. The offset between the curve derived from the unknown-rating model and that from the known-rating model was 0.08. Paired-sample tests conducted using the curve data and the participant's feedback data are shown in Fig. 7.11. Both the unknown-rating model and the known-rating model produced results that did not differ significantly from the participant's feedback data ($p > 0.05$), and the paired correlation analysis indicated a highly significant correlation ($r > 0.99$, $p < 0.05$). Therefore, the model assessment results can be considered both usable and valid.

Competing Interests This study is by "the Fundamental Research Funds for the Central Universities" (NPU Grant No. D5000250035, SEU Grant No. 2242024S30013). Additional support comes from the Northwestern Polytechnical University. The authors have no conflicts of interest to declare that are relevant to the content of this chapter.

References

1. Casali, J. G., & Grenell, J. F. (1990). Noise-attenuating earmuff comfort: A brief review and investigation of band-force, cushion, and wearing-time effects. *Applied Acoustics, 29*(2), 117–138. https://doi.org/10.1016/0003-682X(90)90026-Q
2. Terroir, J., Perrin, N., Wild, P., et al. (2021). Assessing the comfort of earplugs: Development and validation of the French version of the COPROD questionnaire. *Ergonomics, 64*(7), 912–925. https://doi.org/10.1080/00140139.2021.1880027

3. Yan, Y., Fan, H., Li, Y., et al. (2021). Applying wearable technology and a deep learning model to predict occupational physical activities. *Applied Sciences, 11*(20), 1–17. https://doi.org/10.3390/app11209636
4. Yan, Y., Fan, H., Li, Y., et al. (2020). Prediction of occupational physical activities using inertial measurement units and deep learning models. *Proceedings of the Human Factors and Ergonomics Society Annual Meeting, 64*(1), 879–881. https://doi.org/10.1177/1071181320641206
5. Wang, M., Fan, H., Yu, S., et al. (2022). Effects of variations in the tragus expansion angle on physical comfort for in-ear wearables. *Ergonomics, 65*(10), 1–37. https://doi.org/10.1080/00140139.2022.2032377
6. Wheeler, D. E., & Glorig, A. (1956). The industrial hygienist and ear protection. *Noise Control, 2*(1), 45–72. https://doi.org/10.1080/00968205509344015
7. Brown-Rothwell, D. J. (1986). *The comfort of earplugs: Basis for a descriptive model.* University of Southampton.
8. Epps, B. W., & Casali, J. G. (1985). Hearing protection device comfort and user preference: An investigation and evaluation methodology. In *Proceedings of the Human Factors Society Annual Meeting, 29(8)* (pp. 814–818). SAGE Publications. https://doi.org/10.1177/154193128502900821
9. Casali, J. G., Lam, S. T., & Epps, B. W. (1987). Rating and ranking methods for hearing protector wearability. *Sound & Vibration, 21*(12), 10–19.
10. Park, M. Y., & Casali, J. G. (1991). An empirical study of comfort afforded by various hearing protection devices: Laboratory versus field results. *Applied Acoustics, 34*(3), 151–179. https://doi.org/10.1016/0003-682X(91)90082-P
11. Hsu, Y. L., Huang, C. C., Yo, C. Y., et al. (2004). Comfort evaluation of hearing protection. *International Journal of Industrial Ergonomics, 33*(6), 543–551. https://doi.org/10.1016/j.ergon.2004.01.001
12. Ahmadpour, N., Lindgaard, G., Robert, J. M., et al. (2014). The thematic structure of passenger comfort experience and its relationship to the context features in the aircraft cabin. *Ergonomics, 57*(6), 801–815. https://doi.org/10.1080/00140139.2014.899632
13. Laitinen, H., & Poulsen, T. (2008). Questionnaire investigation of musicians' use of hearing protectors, self-reported hearing disorders, and their experience of their working environment. *International Journal of Audiology, 47*(4), 160–168. https://doi.org/10.1080/14992020801886770
14. Bockstael, A., Botteldooren, D., De Bruyne, L., et al. (2012). Personal hearing protection and comfort: Indispensable but not a matter of course. In Proceedings of the European Conference on Noise Control - Euronoise 2012. Prague, Czech Republic: European Acoustics Association (EAA). http://hdl.handle.net/1854/LU-3161933
15. Sviech, P. S., Gonçalves, C. G. O., Morata, T. C., et al. (2013). The evaluation of comfort of the personal hearing protection devices as an intervention for hearing loss prevention. *Revista CEFAC, 15*(5), 1325–1337.
16. Fan, H., Yu, S., Chu, J., et al. (2019). Anthropometric characteristics and product categorization of Chinese auricles for ergonomic design. *International Journal of Industrial Ergonomics, 69*(1), 118–141. https://doi.org/10.1016/j.ergon.2018.11.002
17. Fan, H., Wang, M., Adamson, C. H., et al. (2020). Anthropometric analysis of external acoustic meatus for ergonomic design. *Proceedings of the Human Factors and Ergonomics Society Annual Meeting, 64*(1), 1833–1835. https://doi.org/10.1177/1071181320641441

Appendix

(Data copyright owned by the author Hao Fan.)

Appendix A: Semi-Structured Interview and Structured Questionnaire (Tables A.1, A.2, A.3, A.4, and A.5)

Table A.1 Demographics of participants in the semi-structured interview

Item		N (%) or mean (SD)
Sample size	Total	362 (100%)
	Male	202 (55.80%)
	Female	160 (44.20%)
Age (years)	Total	39.82 (17.71)
	Male	39.98 (17.71)
	Female	39.63 (17.75)
Height (m)	Total	1.69 (0.07)
	Male	1.72 (0.05)
	Female	1.64 (0.06)
Weight (kg)	Total	64.91 (12.53)
	Male	71.46 (10.53)
	Female	56.43 (9.44)
Education	High school or below	63 (17.40%)
	Associate degree	94 (25.97%)
	Bachelor's degree	120 (33.15%)
	Graduate degree (master/ doctor)	85 (23.48%)

(continued)

H. Fan et al., *Engineering Anthropometry of the External Ear for Human-Product Interaction*, https://doi.org/10.1007/978-981-92-2160-8

Table A.1 (continued)

Item		N (%) or mean (SD)
Occupation	Agriculture, forestry, animal husbandry, fishery, water conservancy	8 (2.21%)
	Manufacturing	39 (10.77%)
	Geological survey and exploration	16 (4.42%)
	Construction	22 (6.08%)
	Transportation, postal and telecommunication services	21 (5.80%)
	Commerce, food services, platform delivery, material supply, and logistics	27 (7.46%)
	Real estate management, public utilities, household services, and consulting services	23 (6.35%)
	Health, sports, and social welfare	20 (5.53%)
	Education, culture and arts, broadcasting, new media, and online communication	26 (7.18%)
	Scientific research and technical services	17 (4.70%)
	Finance and insurance	10 (2.76%)
	Government agencies and social organizations	8 (2.21%)
	Students	72 (19.89%)
	Retired individuals	53 (14.64%)

Table A.2 Distribution of participants across age groups in the semi-structured interview

Age group	Total	Male	Female
15–19	66	36	30
20–29	70	42	28
30–39	62	32	30
40–49	50	30	20
50–59	48	26	22
60–69	40	22	18
70+	26	14	12

Table A.3 Demographics of participants in the semi-structured interview (total)

Age group	Total N		Age (years)	Height (m)	Weight (kg)
15–19	66	Mean	18.15	1.67	56.67
		SD	0.77	0.07	10.72
20–29	70	Mean	24.76	1.71	65.69
		SD	2.40	0.07	11.85
30–39	62	Mean	34.95	1.71	67.11
		SD	3.18	0.07	13.14
40–49	50	Mean	44.76	1.71	68.62
		SD	3.11	0.05	11.32
50–59	48	Mean	55.13	1.69	66.35
		SD	2.83	0.07	13.06
60–69	40	Mean	64.60	1.66	66.18
		SD	2.92	0.08	12.00
70+	26	Mean	71.15	1.65	64.69
		SD	1.52	0.06	11.46

Table A.4 Distribution of participants across age groups in the structured questionnaire

Age group	Total	Male	Female
15–19	180	84	96
20–29	201	99	102
30–39	170	86	84
40–49	109	54	55
50–59	89	38	51
60–69	62	35	27
70+	48	22	26

Table A.5 Demographics of participants in the structured questionnaire

Age group	Total N		Age (years)	Height (m)	Weight (kg)
15–19	180	Mean	18.61	1.67	57.44
		SD	0.49	0.07	10.57
20–29	201	Mean	24.74	1.71	66.12
		SD	2.37	0.07	11.80
30–39	170	Mean	34.89	1.70	67.95
		SD	3.14	0.07	13.01
40–49	109	Mean	44.75	1.70	68.87
		SD	3.12	0.05	10.88
50–59	89	Mean	55.18	1.69	67.10
		SD	2.86	0.06	12.89
60–69	62	Mean	64.50	1.67	68.84
		SD	2.92	0.07	11.40
70+	48	Mean	72.17	1.65	67.73
		SD	2.50	0.06	11.13
Total	859	Mean	36.68	1.70	65.40
		SD	16.71	0.07	12.44

Appendix B: Shape-Structure Characterization of the External Ear (Tables B.1, B.2, B.3, B.4, and B.5)

Table B.1 Demographics of participants

Age group	Gender	N	Age (years) Mean ± SD	Height (m) Mean ± SD	Weight (kg) Mean ± SD	BMI (kg/m^2) Mean ± SD
10s	Male	80	17.20 ± 1.28	1.70 ± 0.06	62.25 ± 8.48	21.37 ± 2.22
	Female	80	16.64 ± 1.36	1.61 ± 0.05	47.95 ± 7.59	18.51 ± 2.85
20s	Male	80	24.86 ± 2.32	1.75 ± 0.06	71.50 ± 9.55	23.44 ± 2.78
	Female	80	24.64 ± 2.61	1.66 ± 0.05	56.34 ± 6.83	20.51 ± 2.52
30s	Male	80	34.83 ± 3.00	1.74 ± 0.05	75.91 ± 9.32	24.96 ± 3.01
	Female	80	34.95 ± 3.08	1.66 ± 0.06	58.20 ± 7.49	21.17 ± 2.94
40s	Male	80	44.81 ± 3.02	1.73 ± 0.05	73.29 ± 8.48	24.41 ± 2.68
	Female	80	44.80 ± 2.90	1.66 ± 0.05	59.28 ± 9.12	21.56 ± 3.27
50s	Male	80	55.10 ± 2.85	1.73 ± 0.04	74.95 ± 9.58	25.12 ± 3.20
	Female	80	54.06 ± 2.79	1.63 ± 0.05	59.65 ± 9.38	22.42 ± 3.50
60s	Male	80	64.39 ± 3.00	1.70 ± 0.05	72.06 ± 8.13	24.92 ± 2.86
	Female	80	64.74 ± 3.00	1.60 ± 0.06	57.05 ± 8.91	22.40 ± 3.88
70s	Male	80	73.54 ± 3.32	1.67 ± 0.04	69.93 ± 6.68	25.14 ± 2.45
	Female	80	73.11 ± 2.77	1.60 ± 0.05	56.23 ± 5.90	22.03 ± 2.40
Total	Male	560	44.96 ± 19.39	1.72 ± 0.06	71.41 ± 9.58	24.20 ± 3.03
	Female	560	44.71 ± 19.41	1.63 ± 0.06	56.38 ± 8.74	21.23 ± 3.33

Table B.2 Percentiles of characterization variables for the auricle

No.	Variable	Percentile 5th	10th	25th	50th	75th	90th	95th
1	AL (mm)	66.12	57.35	58.82	61.99	67.53	70.19	72.48
2	AWP (mm)	35.74	30.42	31.51	33.55	35.61	38.20	40.01
3	AWI (mm)	28.22	23.96	25.10	26.62	28.08	30.00	31.56
4	LL (mm)	19.95	15.26	16.13	18.00	19.96	22.04	23.84
5	LW (mm)	19.15	15.02	16.00	17.72	19.04	20.69	22.47
6	EML (mm)	56.43	47.21	48.80	52.57	57.83	60.20	63.04
7	AIA (°)	19.03	12.00	13.36	16.78	19.54	22.00	25.96
8	CL (mm)	29.46	24.97	26.01	27.85	29.41	30.98	33.63
9	CW (mm)	19.62	16.06	16.95	18.26	19.63	20.96	22.13
10	CCL (mm)	18.41	15.42	16.10	17.14	18.49	19.67	20.72
11	CCW (mm)	15.37	12.09	12.71	13.95	15.36	16.73	18.12
12	TL (mm)	18.25	15.33	15.99	17.00	18.31	19.53	20.55
13	TH (mm)	8.37	6.35	6.90	7.59	8.51	9.12	9.89
14	EPS (mm)	16.63	12.72	13.51	15.12	16.53	18.63	19.79
15	EPT (mm)	23.84	18.17	19.35	21.36	24.00	26.30	28.18
16	CD (mm)	16.30	12.95	13.25	14.84	16.10	17.66	19.65
17	CAP (°)	41.91	32.00	35.15	39.00	42.65	46.44	48.90
18	CAO (°)	57.36	40.14	43.00	48.25	56.36	66.00	74.22

Table B.3 Percentiles of characterization variables for the external acoustic meatus

No.	Variable	Percentile 5th	10th	25th	50th	75th	90th	95th
19	EL (mm)	13.09	13.57	14.56	15.30	16.22	17.26	17.69
20	EW (mm)	6.89	7.39	8.11	8.74	9.14	9.70	9.95
21	EC (mm)	34.70	36.45	38.25	40.21	41.86	44.48	45.54
22	EA (mm^2)	77.94	83.85	94.51	105.77	114.84	124.47	130.41
23	FBL (mm)	7.52	8.15	8.84	9.73	10.59	11.38	11.70
24	FBW (mm)	6.05	6.40	7.18	7.76	8.23	8.89	9.20
25	FBC (mm)	23.03	24.07	26.29	28.02	30.00	31.71	32.49
26	FBA (mm^2)	38.74	42.61	51.80	58.93	66.37	72.27	75.50
27	SBL (mm)	6.49	7.05	7.53	7.96	8.34	8.82	9.14
28	SBW (mm)	6.07	6.20	6.65	7.34	7.77	8.34	8.85
29	SBC (mm)	21.11	21.87	23.07	24.20	25.38	26.89	27.91
30	SBA (mm^2)	33.86	36.62	40.50	45.09	49.81	54.39	58.73
31	SD-I (mm)	5.47	5.85	6.50	7.35	8.53	9.29	9.83
32	SD-II-F (mm)	5.24	5.51	6.16	7.03	8.17	8.85	9.31
33	SD-II-E (mm)	10.50	10.95	11.91	13.42	15.04	16.25	16.83
34	ED-C (mm)	11.32	12.09	13.30	14.17	15.04	15.83	16.30
35	ED-P (mm)	8.08	9.55	10.33	10.92	11.74	12.67	13.49
36	ACE (mm)	18.84	20.48	23.38	25.25	26.59	28.19	29.48
37	AS-I (mm)	98.23	100.46	103.88	107.91	113.47	117.44	118.51
38	ES-I (mm)	8.54	10.10	11.61	14.15	17.43	23.91	26.74
39	AS-II (mm)	119.81	123.12	128.37	133.78	139.48	145.36	148.67
40	ES-II (mm)	21.35	22.91	25.44	28.89	33.89	39.10	42.90

Table B.4 Multiple comparisons of characterization variables for the auricle (age groups)

No.	Variable	Age group						
		10s (ear, N = 320)	20s (ear, N = 320)	30s (ear, N = 320)	40s (ear, N = 320)	50s (ear, N = 320)	60s (ear, N = 320)	70s (ear, N = 320)
		Mean ± SD	Mean ± SD	Mean ± SD	Mean ± SD	Mean ± SD	Mean ± SD	Mean ± SD
1	AL (mm)	$59.00^{a} \pm 3.76$	$62.33^{b} \pm 3.34$	$64.55^{c} \pm 4.15$	$65.57^{d} \pm 3.88$	$69.43^{e} \pm 2.51$	$70.34^{f} \pm 2.04$	$72.12^{g} \pm 2.70$
2	AWP (mm)	$32.51^{a} \pm 2.40$	$34.09^{b} \pm 2.37$	$34.71^{c} \pm 2.78$	$34.90^{c} \pm 2.34$	$37.21^{d} \pm 2.24$	$37.68^{d} \pm 2.93$	$39.40^{e} \pm 3.42$
3	AWI (mm)	$26.27^{a} \pm 2.10$	$26.84^{b} \pm 2.06$	$27.58^{c} \pm 2.14$	$28.00^{c} \pm 2.40$	$28.70^{d} \pm 2.19$	$29.51^{e} \pm 2.66$	$30.40^{f} \pm 2.05$
4	LL (mm)	$17.09^{a} \pm 2.32$	$18.86^{b} \pm 2.55$	$19.13^{b} \pm 2.40$	$19.38^{b} \pm 2.53$	$21.06^{c} \pm 2.48$	$21.65^{d} \pm 2.30$	$22.61^{e} \pm 2.19$
5	LW (mm)	$18.12^{a} \pm 2.11$	$18.30^{a} \pm 2.15$	$18.30^{a} \pm 2.09$	$18.88^{b} \pm 1.87$	$19.77^{c} \pm 2.91$	$20.18^{c} \pm 2.64$	$20.71^{d} \pm 1.91$
6	EML (mm)	$50.01^{a} \pm 4.00$	$54.15^{b} \pm 4.19$	$55.81^{c} \pm 4.71$	$55.87^{c} \pm 4.51$	$58.63^{d} \pm 3.71$	$59.09^{d} \pm 3.44$	$62.20^{e} \pm 2.87$
7	AIA (°)	$18.19^{a} \pm 3.58$	$19.10^{b,c} \pm 5.62$	$19.48^{c} \pm 4.06$	$18.83^{b} \pm 4.71$	$18.23^{a} \pm 3.44$	$20.56^{d} \pm 4.90$	$18.72^{b} \pm 4.07$
8	CL (mm)	$26.91^{a} \pm 2.63$	$28.60^{b} \pm 1.80$	$28.66^{b} \pm 3.30$	$29.35^{c} \pm 2.03$	$29.95^{d} \pm 2.80$	$30.45^{d} \pm 2.46$	$32.58^{e} \pm 3.60$
9	CW (mm)	$18.43^{a} \pm 2.17$	$19.01^{b} \pm 1.77$	$19.24^{b} \pm 1.73$	$19.67^{c} \pm 1.92$	$20.14^{d} \pm 1.71$	$20.58^{d} \pm 2.54$	$20.48^{d} \pm 2.35$
10	CCL (mm)	$16.91^{a} \pm 1.96$	$18.05^{b} \pm 1.62$	$18.22^{b} \pm 1.48$	$18.73^{c} \pm 1.56$	$18.88^{c} \pm 1.61$	$18.94^{c} \pm 1.76$	$18.81^{c} \pm 2.17$
11	CCW (mm)	$14.43^{a} \pm 2.09$	$14.75^{b} \pm 1.92$	$14.77^{b} \pm 1.74$	$15.23^{c} \pm 1.97$	$15.78^{d} \pm 1.80$	$16.45^{e} \pm 2.50$	$16.28^{e} \pm 2.38$
12	TL (mm)	$16.77^{a} \pm 1.95$	$18.03^{b} \pm 1.60$	$18.17^{b} \pm 1.51$	$18.59^{c} \pm 1.61$	$18.75^{c} \pm 1.58$	$18.94^{c} \pm 1.75$	$18.68^{c} \pm 2.12$
13	TH (mm)	$7.92^{a} \pm 1.03$	$8.44^{b,c} \pm 1.50$	$8.80^{c} \pm 1.09$	$8.65^{c} \pm 1.02$	$8.85^{c} \pm 1.12$	$8.24^{b} \pm 1.29$	$8.12^{a,b} \pm 1.30$
14	EPS (mm)	$15.88^{a} \pm 2.10$	$16.14^{a} \pm 2.62$	$17.57^{c} \pm 2.23$	$17.20^{c} \pm 2.50$	$17.02^{c} \pm 2.31$	$16.53^{b} \pm 2.45$	$16.52^{b} \pm 2.43$
15	EPT (mm)	$23.16^{a} \pm 3.42$	$24.00^{b} \pm 3.12$	$24.29^{b} \pm 3.35$	$25.08^{c} \pm 4.04$	$24.49^{b} \pm 3.36$	$23.49^{b} \pm 2.59$	$24.35^{b} \pm 3.26$
16	CD (mm)	$15.53^{a} \pm 2.06$	$16.37^{b,c} \pm 2.32$	$16.36^{b} \pm 2.25$	$17.24^{d} \pm 2.66$	$16.79^{c} \pm 2.50$	$16.28^{b} \pm 2.65$	$15.83^{a,b} \pm 1.97$
17	CAP (°)	$41.78^{a} \pm 4.54$	$41.83^{a} \pm 5.23$	$44.41^{c} \pm 4.10$	$43.43^{c} \pm 5.12$	$43.14^{c} \pm 4.85$	$42.53^{b} \pm 5.65$	$42.51^{b} \pm 5.55$
18	CAO (°)	$57.23^{b} \pm 11.67$	$59.85^{c} \pm 10.93$	$60.17^{c} \pm 12.74$	$60.76^{d} \pm 12.14$	$59.69^{c} \pm 13.20$	$57.60^{b} \pm 10.16$	$54.60^{a} \pm 12.19$

a, b, c, d, e, f, g: multiple comparison results of Tukey test, the same superscript on mean values indicated no significant mean differences between age groups.
$\alpha = 0.05$
N: number of ears

Table B.5 Multiple comparisons of characterization variables for the external acoustic meatus (age groups)

No.	Variable	Age group 10s (ear, N = 320)	20s (ear, N = 320)	30s (ear, N = 320)	40s (ear, N = 320)	50s (ear, N = 320)	60s (ear, N = 320)	70s (ear, N = 320)
		Mean ± SD	Mean ± SD	Mean ± SD	Mean ± SD	Mean ± SD	Mean ± SD	Mean ± SD
19	EL (mm)	$13.76^{a} \pm 1.28$	$14.63^{b} \pm 0.94$	$14.98^{c} \pm 0.70$	$15.34^{d} \pm 1.06$	$15.52^{d} \pm 0.94$	$16.61^{e} \pm 0.91$	$16.64^{e} \pm 0.93$
20	EW (mm)	$7.83^{a} \pm 0.73$	$8.51^{b} \pm 0.55$	$8.91^{c} \pm 0.69$	$9.01^{c} \pm 0.84$	$8.97^{c} \pm 0.76$	$8.68^{b} \pm 0.86$	$8.45^{b} \pm 1.09$
21	EC (mm)	$36.51^{a} \pm 2.76$	$38.95^{b} \pm 1.98$	$39.82^{c} \pm 1.87$	$40.25^{c} \pm 2.58$	$40.46^{c} \pm 2.26$	$42.01^{d} \pm 2.86$	$43.18^{e} \pm 3.18$
22	EA (mm^2)	$85.30^{a} \pm 12.17$	$98.58^{b} \pm 9.07$	$104.5^{c} \pm 9.38$	$108.31^{c} \pm 14.15$	$110.71^{d} \pm 12.93$	$113.95^{d} \pm 12.04$	$112.28^{d} \pm 16.88$
23	FBL (mm)	$8.37^{a} \pm 1.00$	$8.73^{b} \pm 0.98$	$9.75^{c} \pm 0.71$	$9.84^{c} \pm 0.46$	$10.79^{d} \pm 0.94$	$10.31^{c} \pm 1.15$	$10.13^{c} \pm 0.89$
24	FBW (mm)	$7.05^{a} \pm 0.97$	$7.42^{b} \pm 0.95$	$7.90^{c} \pm 0.8$	$8.05^{d} \pm 0.72$	$7.97^{d} \pm 0.72$	$7.75^{c} \pm 0.86$	$7.71^{c} \pm 0.90$
25	FBC (mm)	$24.80^{a} \pm 2.45$	$25.89^{b} \pm 2.17$	$28.38^{c} \pm 1.64$	$28.70^{c} \pm 1.35$	$30.38^{d} \pm 2.39$	$28.91^{c} \pm 2.64$	$28.85^{c} \pm 2.06$
26	FBA (mm^2)	$46.65^{a} \pm 9.48$	$50.79^{b} \pm 8.66$	$60.46^{c} \pm 7.27$	$61.61^{c} \pm 6.07$	$67.19^{d} \pm 9.50$	$62.11^{c} \pm 10.35$	$60.94^{c} \pm 9.86$
27	SBL (mm)	$7.54^{a} \pm 0.82$	$7.89^{b} \pm 0.71$	$8.04^{b} \pm 0.83$	$8.13^{b} \pm 0.60$	$8.05^{b} \pm 0.72$	$8.00^{b} \pm 0.75$	$7.92^{b} \pm 0.74$
28	SBW (mm)	$6.40^{a} \pm 0.61$	$6.94^{b} \pm 0.62$	$7.35^{c} \pm 0.55$	$7.89^{d} \pm 0.92$	$7.53^{c} \pm 0.77$	$7.53^{c} \pm 0.78$	$7.45^{c} \pm 0.83$
29	SBC (mm)	$22.38^{a} \pm 1.81$	$23.63^{b} \pm 1.60$	$24.58^{c} \pm 1.90$	$25.17^{d} \pm 1.62$	$24.68^{c,d} \pm 1.51$	$24.77^{c,d} \pm 2.15$	$24.59^{c} \pm 2.41$
30	SBA (mm^2)	$37.85^{a} \pm 5.62$	$42.84^{b} \pm 5.61$	$46.61^{c} \pm 6.15$	$49.56^{d} \pm 7.13$	$47.30^{c} \pm 6.15$	$46.95^{c} \pm 7.45$	$46.37^{c} \pm 6.60$
31	SD-I (mm)	$6.13^{a} \pm 0.70$	$6.92^{b} \pm 1.02$	$6.92^{b} \pm 1.05$	$7.74^{c} \pm 1.18$	$8.05^{c} \pm 1.42$	$8.32^{d} \pm 1.26$	$8.50^{d} \pm 0.88$
32	SD-II-F (mm)	$6.38^{a} \pm 0.91$	$6.99^{b} \pm 1.62$	$7.14^{b,c} \pm 1.31$	$7.26^{b,c} \pm 1.39$	$7.3^{b,c} \pm 1.20$	$7.42^{c} \pm 0.98$	$7.90^{d} \pm 1.20$
33	SD-II-E (mm)	$11.41^{a} \pm 1.03$	$12.76^{b} \pm 1.95$	$12.90^{b} \pm 1.66$	$13.77^{c} \pm 1.88$	$14.01^{c} \pm 1.85$	$14.54^{d} \pm 1.63$	$15.30^{e} \pm 1.45$
34	ED-C (mm)	$11.93^{a} \pm 1.08$	$14.07^{b} \pm 1.05$	$13.85^{b} \pm 0.85$	$14.06^{b} \pm 0.90$	$14.97^{c} \pm 1.12$	$14.98^{c} \pm 0.95$	$14.70^{c} \pm 1.41$
35	ED-P (mm)	$9.02^{a} \pm 1.25$	$10.56^{b} \pm 0.83$	$10.91^{c} \pm 0.69$	$11.14^{c} \pm 0.78$	$11.65^{d} \pm 0.76$	$11.79^{d} \pm 1.15$	$11.80^{d} \pm 1.35$
36	ACE (mm)	$20.90^{a} \pm 2.72$	$23.13^{b} \pm 2.61$	$24.30^{c} \pm 1.78$	$24.80^{c} \pm 2.06$	$26.08^{d} \pm 1.50$	$26.95^{e} \pm 1.86$	$27.75^{f} \pm 2.25$
37	AS-I (mm)	$105.25^{a} \pm 6.77$	$106.48^{a,b} \pm 6.16$	$106.98^{a,b} \pm 6.84$	$107.80^{b} \pm 5.42$	$109.81^{c} \pm 5.83$	$111.20^{c} \pm 5.34$	$111.31^{c} \pm 5.45$
38	ES-I (mm)	$12.10^{a} \pm 2.60$	$13.00^{a} \pm 4.53$	$14.20^{b} \pm 5.51$	$15.05^{b} \pm 5.43$	$17.05^{c} \pm 5.39$	$16.59^{c} \pm 5.41$	$18.70^{d} \pm 5.45$

(continued)

Table B.5 (continued)

		Age group						
		10s (ear, N = 320)	20s (ear, N = 320)	30s (ear, N = 320)	40s (ear, N = 320)	50s (ear, N = 320)	60s (ear, N = 320)	70s (ear, N = 320)
No.	Variable	Mean ± SD	Mean ± SD	Mean ± SD	Mean ± SD	Mean ± SD	Mean ± SD	Mean ± SD
39	AS-II (mm)	$132.15^{a} \pm 8.43$	$132.25^{a} \pm 8.41$	$133.13^{a,b} \pm 8.10$	$133.21^{a,b} \pm 6.5$	$132.77^{a,b} \pm 8.95$	$135.77^{b} \pm 10.37$	$138.24^{c} \pm 9.43$
40	ES-II (mm)	$26.25^{a} \pm 3.61$	$27.87^{a,b} \pm 5.02$	$28.30^{b} \pm 7.04$	$33.45^{d} \pm 7.44$	$34.37^{d} \pm 7.14$	$30.80^{c} \pm 5.30$	$30.21^{c} \pm 6.02$

a, b, c, d, e, f: Multiple Comparison Results of Tukey Test, the same superscript on mean values indicated no significant mean differences between age groups.
$\alpha = 0.05$
N: number of ears

Appendix C: Rapid Estimation Models of the External Acoustic Meatus Cross-Section (Table C.1, C.2, and C.3)

Table C.1 Difference analysis of auricular linear variables for wearing comfort

Variable	Comfort rating	Description	Sexual difference (independent-samples)		Bilateral difference (paired-samples)		Age-group difference (ANOVA)	
			Mean difference (%)	p	Mean difference (%)	p	df	p
AL	−1	Slight discomfort	0.16	0.10	0.01	0.91	3	0.06
	−2	Moderate discomfort	−0.02	0.80	−0.18	0.01	3	0.10
	−5	Complete discomfort	−0.15	0.13	0.004	0.97	3	0.83
AWP	−1	Slight discomfort	0.10	0.19	−0.10	0.19	3	0.54
	−2	Moderate discomfort	0.03	0.67	−0.04	0.58	3	0.73
	−5	Complete discomfort	−0.10	0.22	0.11	0.18	3	0.60
AWI	−1	Slight discomfort	−0.004	0.95	−0.02	0.77	3	0.38
	−2	Moderate discomfort	0.003	0.97	−0.15	0.14	3	0.32
	−5	Complete discomfort	0.12	0.25	−0.07	0.52	3	0.18
EML	−1	Slight discomfort	−0.02	0.78	0.09	0.16	3	0.95
	−2	Moderate discomfort	0.02	0.82	−0.13	0.09	3	0.72
	−5	Complete discomfort	0.10	0.28	0.18	0.06	3	0.30

Table C.2 Wearing comfort thresholds and exposure limits of deformation ratios (individual threshold)

Comfort rating	Description	Variable			
		AL (%)	AWP (%)	AWI (%)	EML (%)
0	No discomfort	0.00	0.00	0.00	0.00
−0.5	Just-noticeable discomfort	7.04	5.06	5.24	4.12
−1	Slight discomfort	8.84	7.20	7.50	5.79
−1.5	Slight discomfort	10.03	8.81	9.20	7.11
−2	Moderate discomfort	11.00	10.20	10.63	8.28
−2.5	Moderate discomfort	11.90	11.48	11.91	9.38
−3	High discomfort	12.74	12.71	13.10	10.46
−3.5	High discomfort	13.59	13.86	14.23	11.58
−4	Extreme discomfort	14.48	15.02	15.30	12.72
−4.5	Extreme discomfort	15.44	16.16	16.35	13.94
−5 (limit)	Complete discomfort	16.51	17.33	17.36	15.26

Table C.3 Wearing comfort thresholds of deformation for mean size (individual threshold)

Comfort rating	Reference value	Comfort rating										
		0	−0.5	−1	−1.5	−2	−2.5	−3	−3.5	−4	−4.5	−5 (限值)
AL (mm)	Female	64.92	69.49	70.66	71.43	72.06	72.64	73.19	73.75	74.32	74.94	64.92
AWP (mm)	Female	35.02	36.79	37.54	38.10	38.59	39.04	39.47	39.87	40.28	40.68	35.02
AWI (mm)	Female	27.52	28.96	29.58	30.05	30.45	30.80	31.12	31.44	31.73	32.02	27.52
EML (mm)	Total	56.43	58.76	59.70	60.44	61.10	61.72	62.33	62.96	63.61	64.30	56.43

Appendix D: Auricle-EAM Transitional Structure and Its Expansion (Tables D.1 and D.2)

Table D.1 Percentiles of the measurement variable for the auricle-EAM transitional structure

No.	Variable	Percentile						
		5th	10th	25th	50th	75th	90th	95th
41	ATEMC (°)	22.71	24.47	26.85	30.42	36.31	39.86	42.11

Table D.2 Multiple comparisons of the measurement variable for the auricle-EAM transitional structure (by age groups)

No.	Variable	Age group						
		10s (ear, N = 320)	20s (ear, N = 320)	30s (ear, N = 320)	40s (ear, N = 320)	50s (ear, N = 320)	60s (ear, N = 320)	70s (ear, N = 320)
		Mean ± SD	Mean ± SD	Mean ± SD	Mean ± SD	Mean ± SD	Mean ± SD	Mean ± SD
41	ATEMC (°)	$27.20^{a} \pm 3.02$	$28.61^{a} \pm 4.55$	$30.47^{b} \pm 5.35$	$32.48^{c} \pm 6.40$	$33.05^{c} \pm 4.82$	$34.90^{d} \pm 4.05$	$35.10^{d} \pm 7.40$

a, b, c, d: Multiple Comparison Results of Tukey Test, the same superscript on mean values indicated no significant mean differences between age groups.
$\alpha = 0.05$
N: number of ears

Appendix E: Comfort Loss (Tables E.1 and E.2)

Table E.1 Questionnaire scale for wearing comfort loss process of motivation

Ear-related wearable comfort questionnaire
Section II: Assessment of wearing comfort loss based on usage motivation
Please evaluate the changes in wearing comfort loss under the following usage scenarios/contexts according to your personal experience or daily usage. Indicate the corresponding score in the parentheses after each time option Conditions for the validity of each usage scenario/context 1. The device you used is considered by you as the most suitable (i.e., optimal comfort product) 2. You were in a relaxed physical and mental state during the given scenario or context 3 You are very familiar with the entire process of the given scenario/context event 4. Sweating is not the reason for your comfort loss If you do not have long-term experience in a given scenario/context under the above conditions, you may leave the question unanswered For each question, two device-assessment items are provided. Please specify which type of product you usually use to complete the given scenario/context event. If you typically use only one type, fill in only one row. Device types include Invasive devices (which cause tragus expansion during fitting; comfort loss scale = 0 to −10) Non-invasive devices (comfort loss scale = 0 to −5) You may use 0.5 increments (e.g., −0.5, −1.5, etc.) When assessing comfort loss, please refer to your daily single-use duration as the basis. For example, if during jogging you remove or adjust the device after 90 min due to intolerable pressure or expansion of the external ear, then the comfort loss at 90 min should be marked as −5. If you have never experienced a −5 loss within the given time options, you do not need to mark the final option as −5. If your actual usage time is shorter than the given options, only fill in up to your actual duration
1. Scenario/context 1: Household Event: Cooking • Device item 1 15 min (); 30 min (); 45 min (); 60 min () 75 min (); 90 min (); 105 min (); 120 min () • Device item 2: 15 min (); 30 min (); 45 min (); 60 min (); 75 min (); 90 min (); 105 min (); 120 min ()
2. Scenario/context 2: Exercise Event: Brisk walking • Device item 1: 15 min (); 30 min (); 45 min (); 60 min () 75 min (); 90 min (); 105 min (); 120 min () • Device item 2: 15 min (); 30 min (); 45 min (); 60 min (); 75 min (); 90 min (); 105 min (); 120 min ()
3. Scenario/context 2: Exercise Event: Jogging • Device item 1 15 min (); 30 min (); 45 min (); 60 min () 75 min (); 90 min (); 105 min (); 120 min () • Device item 2 15 min (); 30 min (); 45 min (); 60 min () 75 min (); 90 min (); 105 min (); 120 min ()

(continued)

Table E.1 (continued)

4. Scenario/context 3: Office
Event: Sedentary work
- Device item 1
  15 min (); 30 min (); 45 min (); 60 min ()
  75 min (); 90 min (); 105 min (); 120 min ()
  135 min (); 150 min (); 165 min (); 180 min ()
  195 min (); 210 min (); 225 min (); 240 min ()
- Device item 2:
  15 min (); 30 min (); 45 min (); 60 min ()
  75 min (); 90 min (); 105 min (); 120 min ()
  135 min (); 150 min (); 165 min (); 180 min ()
  195 min (); 210 min (); 225 min (); 240 min ()

Table E.2 Statistics for wearing comfort loss process of motivation

Duration of use (mins)	Comfort loss														
	Cooking (Invasive device)			Brisk walking (invasive device)			Jogging (invasive device)			Sedentary work 1 (invasive device)			Sedentary work 2 (non-invasive device)		
	N	Mean	SD	N	Mean	SD	N	Mean	SD	N	Mean	SD	N	Mean	SD
15	365	−0.01	0.07	443	−0.17	0.27	420	−0.17	0.25	322	−0.09	0.22	306	−0.02	0.10
30	365	−0.04	0.13	443	−0.40	0.20	420	−0.34	0.23	322	−0.53	0.37	306	−0.12	0.21
45	365	−0.15	0.26	443	−0.56	0.18	420	−0.76	0.40	322	−0.91	0.60	306	−0.38	0.22
60	365	−0.54	0.40	443	−0.89	0.29	420	−1.38	0.77	322	−1.61	1.05	306	−0.62	0.25
75	365	−0.80	0.45	442	−1.61	0.51	420	−2.05	0.90	322	−2.54	1.23	306	−0.88	0.45
90	364	−1.33	0.71	382	−2.20	0.36	420	−2.74	0.93	322	−3.63	1.34	306	−1.28	0.58
105	355	−1.76	0.53	327	−3.15	0.37	415	−3.49	0.86	322	−4.44	1.20	306	−1.70	0.60
120	282	−2.14	0.38	287	−4.05	0.35	402	−4.35	0.64	322	−5.13	1.18	306	−2.14	0.60
135										322	−5.78	1.20	306	−2.51	0.60
150										322	−6.45	1.26	306	−2.82	0.60
165										322	−7.18	1.38	306	−3.19	0.66
180										303	−7.80	1.40	306	−3.56	0.71
195										242	−8.07	1.20	288	−3.89	0.72
210										205	−8.58	1.12	231	−4.03	0.68
225										146	−9.09	1.04	189	−4.28	0.57
240										63	−9.22	0.92	127	−4.56	0.51

Key Laboratory of Industrial Design and Ergonomics, Ministry of Industry and Information Technology, ChinaThis work was supported by Key Laboratory of Industrial Design and Ergonomics, Ministry of Industry and Information Technology, China.

MIX
Papier aus verantwortungsvollen Quellen
Paper from responsible sources
FSC® C105338

If you have any concerns about our products,
you can contact us on
ProductSafety@springernature.com

In case Publisher is established outside the EU,
the EU authorized representative is:
Springer Nature Customer Service Center GmbH
Europaplatz 3, 69115 Heidelberg, Germany

Printed by Libri Plureos GmbH
in Hamburg, Germany